W0246646

IMAGES AND CONTEXTS

IMAGES AND CONTEXTS

The Historiography of Science and Modernity in India

DHRUV RAINA

OXFORD
UNIVERSITY PRESS

OXFORD
UNIVERSITY PRESS

Oxford University Press is a department of the University of Oxford.
It furthers the University's objective of excellence in research, scholarship,
and education by publishing worldwide. Oxford is a registered trademark of
Oxford University Press in the UK and in certain other countries

Published in India by
Oxford University Press
2/11 Ground Floor, Ansari Road, Daryaganj, New Delhi 110 002, India

First published 2003
Oxford India Paperbacks 2010

ISBN-13 (print edition): 978-0-19-806880-8
ISBN-10 (print edition): 0-19-806880-8

ISBN-13 (eBook): 978-0-19-908829-4
ISBN-10 (eBook): 0-19-908829-2

I have serious reasons to believe that the little prince's planet of origin was the asteroid known as B–612. This asteroid has only been observed once through a telescope by a Turkish astronomer in 1909. At the time, he organised a great demonstration of his discovery at an International Astronomical Congress. But because of his Turkish attire, nobody believed him. Grown-ups are like that.

Fortunately, for the reputation of Asteroid B–612, however, a Turkish dictator imposed European costume upon his subjects under pain of death. So the astronomer repeated his demonstration in 1920, dressed in an elegant suit. And this time, everybody was convinced.

Antoine De Saint Exupery
The Little Prince
(Tr. Irene Testot-Ferry), Hermes Inc. 1999.

Contents

Acknowledgements

The essays that have been put together in this book were written and published between 1993 and 2000. While they do not in any sense constitute a comprehensive study in the historiography of sciences of India they address certain issues that have drawn my attention. These essays have benefited from innumerable exchanges with scholars from a variety of disciplines and commitments. Most of these essays were first presented at various workshops and seminars before they acquired the form they have; and but for a few I have not done much to alter or revise them as will be evident to the reader. I am beholden to all those who have patiently sat through some of these workshops and seminars, agreed or disagreed with these presentations, and furthered my understanding. My colleague and collaborator S. Irfan Habib has been a sympathetic discussant for each of these essays. National Institute of Science, Technology and Development Studies, till quite recently, was home to a number of scholars working on the history of science during the period of colonial rule, and amongst them I take this opportunity to thank Pratik Chakravarty, Dhirendra Dangwal, Amitabh Ghosh, Deepak Kumar, and Harish Khatri Naraindass for the heated discussions. This give-and-take has taught me that we are deeply indebted to the views of those with whom we most strongly disagree. Shiv Visvanathan encouraged me to finish this book. The second of these essays was prompted by a request from Aant Elzinga, and he has borne the brunt of having to read the rest that followed. But for the suggestions of Arion Roşu and Santimoy Chatterjee and the support of the Maison de Sciences de l'Homme the third of these chapters would not have been possible. And throughout all this Bob Anderson, Sentil Babu, Prajit K. Basu, Gregory Blue, Karine Chemla, Subrata Dasgupta, Heloisa Dominguez, Steve Fuller, Merle Jacob, Catherine Jami, Helen Longino, Michel

Paty, Jacques Pouchepadass, Patrick Petitjean, Kapil Raj, Abha Sur, Roshdi Rashed, Romila Thapar, and T.V. Venkateswaran have commented on one or several of the essays appearing here. The tedious task of classifying the IJHS publications and tabulating them was undertaken by Sunil Sharma, a student trainee from BITS, Pilani. The author is indebted to his 'Bibliography of IJHS Publications (1966–1994)', NISTADS Report, 1995. This bibliography forms the basis for the analysis in Chapter 5.

The sources for published essays appearing in this volume are as follows:

Raina, Dhruv. 1996. 'Reconfiguring the Centre! On the Structure of Scientific Exchanges in Colonial India', *Minerva* 34, pp. 161–76.

—— 1998. 'Historiographic Concerns Underlying Indian Journal of History of Science: A Bibliometric Inference', *Economic and Political Weekly* 33(8), pp. 407–14.

—— 'Evolving Perspectives on Science and History: A Chronicle of Modern India's Scientific Enchantment and Disenchantment (1850–1980)', *Social Epistemology* 11(1), pp. 3–24.

—— 1997. 'The Early Years of P.C. Ray: The Inauguration of the School of Chemistry and the Social History of Science', *Science Technology & Society* 2(1), pp. 1–40. [Copyright © Society for the Promotion of S&T Studies, New Delhi. All rights reserved. Reproduced with the permission of the copyright holders and the publishers, Sage Publications India Pvt. Ltd, New Delhi.]

Raina, Dhruv and Habib, S. Irfan. 1999. 'The Missing Picture: The Non-emergence of a Needhamian History of Sciences of India', in S. Irfan Habib and Dhruv Raina (eds), *Situating the History of Sciences: Dialogues with Joseph Needham*. New Delhi: Oxford University Press, pp. 279–302.

Raina, Dhruv. 1999. 'From West to Non-West?: Basalla's Three Stage Model Revisited', *Science as Culture* 8(4), pp. 497–516.

—— 2000. 'The Present in the Past: Trajectories for the Social History of Science', in Romila Thapar (ed.), *India: Another Millennium?* New Delhi: Viking, pp. 17–35.

I thank Oxford University Press and Penguin, and the publishers of *Minerva*, *Economic and Political Weekly*, *Social Epistemology* (http:\\www.tandf.co.uk), *Science, Technology and Society*, and *Science as Culture* (http:\\www.tandf.co.uk) for permission to publish these essays. The editors of Oxford University Press were very helpful and patiently put up with the repeated postponement of deadlines. I take a leaf out from the book of my philosopher and friend Santanagopal

Rajagoplan, who informs me that in the grand Indian tradition we thank those who helped the most, last. And so I dedicate this book to my wife Rajeswari who has been party to my reflexive turn to the study of the history of history of science.

List of Tables and Figures

TABLES

FIGURES

1

Introduction

The study of the knowledge forms of non-Western societies is a rapidly emerging research field that could have a lasting impact on the disciplinary history of science. Long marginalized by positivist and triumphalist histories of science, the history of non-Western knowledge forms presents opportunities for rethinking the discipline. This renewal is a product of the study of these knowledge forms as well a consequence of the changes within the discipline itself. The last three decades of the twentieth century witnessed a critical turn towards the theory of history and an examination of the representation and historiography of the Orient. On this count, the now canonized works of Said (1978, 1994), Bernal (1987), Inden (1990), and others mark an undeniable departure. A similar turn was witnessed in the sociology of scientific knowledge, and was reflected in the preoccupation of researchers with the reflexivity issue in science studies, the need to situate the analyst's account, and to contextualize it just as the analyst was busy putting the object of her study, be it a scientific research programme or a scientific controversy, or the emergence of national scientific research system, in context (Barnes and Edge 1982; Pestre 1995). However, the focus of such a theoretical renewal as far as the south Asian region is concerned has largely been in the domain of literary, anthropological, and historical production and not so much on the production of historical knowledge of the natural sciences. Social theorists of science have not turned their reflexive torch to locate the historiography of the natural sciences of India. This situation needs addressing, and I have attempted in a very modest way to engage with the problematic. Over the last decade I have been researching the subject, given that my location is that of a historian and philosopher of science, and have attempted a social epistemology of the science and history archive on India.

The subject essays that follow are not so much about how the sciences of the Orient were represented in the discourse produced by the Occident, but more importantly how Indian scientists and historians of science engaged with the sciences of India. What did they inherit from the Occidental discourse about the Orient and where did they depart from the former? What did the globalization of modern science mean for these scientist-historians and how did this frame their historical encounter with the sciences of India? Was the production of the history of sciences a cultural activity intended to legitimate the assimilation of modern science within the rubric of the modern Indian state—both colonial and post-colonial? A second set of concerns relates to how the nationalist movement and the emerging bonds between science and the modern nation state resonated with policy discourse and academic representations and images of science. From the Orientalist period, through the nationalist period, to the period of Nehruvian socialism in India, the forces of decolonization sought to challenge a hegemonic conception of the history of science, such that the history of sciences of the non-West would not be subordinate texts to the mainstream discourse of the history of Western science. Colonialism was an important milestone in this history, and the process of decolonization prompted a number of reverse commentaries. Situating these reverse commentaries not so much historically but metatheoretically is what brings these essays thematically together.

There is an interesting discussion in Hobsbawm's *A Short History of the Twentieth Century* about the emergence of a range of reflexive discourses, for example, a whole genre of contemporary cinema targeted at university educated audiences capable of deciphering the metatheory of cinema, or the emergence of the university novel as a literary genre (Hobsbawm 1993). With the rapid dissemination of literacy in Europe in the post-World War II years, academic interest turned to the social production of discourses, the sites of production of such discourses, their audiences, publics, and readerships. This reflexive turn was marked by an intellectual disenchantment that was at once both ideological and cultural: in the former case, the rise of social-ism and the emergence of the third world from the shadow of late 19th-century colonialism altered the political power relations in the developed world. Culturally this generated a diminution of the importance of Europe as the site of knowledge production and generated an increasing interest in the construction of knowledge. The dialectic of enlightenment had now moved towards the critique of modernity and its most potent symbol, science.

The attempt to situate the history of sciences is evident in the critiques of the early studies of the work of Joseph Needham, but acquires its most explicit formulation in Loren Graham's attempt to put Boris Hessen's historiography in its socio-political context.[1] As Graham (1985) put it, Hessen appeared to be telling the Soviet authorities that if they could label Einstein's science bourgeois, then he could very well trace its seeds to Newton's project—the very Newton who had been lionized as one of the epitomes of materialist thought. Consequently, the methods the historian employed to contextualize the efforts of the scientists could be turned upon the historian's reconstructions. Such an enterprise could run the danger of relativizing scientific activity. But this is not a necessary outcome if scientific theories are seen as 'realistic maps or explanations of the real world', that encode 'cultural and linguistic categories and cultural values'. Different versions of such a constructivist position have been proposed, two of them being realistic constructivism and constructive realism (Hess 1997: 35–7).[2] Of late there has been a personalization of historical scholarship evident in the emergence of social history. But it should nevertheless be pointed out that the waves of postmodernist relativism broke last of all upon the discipline of history, since most historians were committed to the notion that 'they were building not upon sand but upon the solid ground of fact' (Fox-Genovese and Lasch-Quinn 1999: xiv). The idea of reflexivity extends constructivism to the very theories produced by social scientists about science in society. This means that not only cultural practices (Pickering 1992), but also non-evidential considerations enter into the production of social theories of science.

The social turn in science studies has struck a resonant chord among third-world scholars who were grappling with related issues. The depriveleging of the sacred cow of 'scientific rationality' in its Western avatar provided an opening for critically re-evaluating indigenous knowledge systems. Second, social movements within these societies resisting 'steam roller development' were re-examining processes underlying the cultural appropriation of a new knowledge form and its concomitant practices. On the other hand there were movements attempting to advance the place of science in social transformation. In either case the central political agenda was that of democratization. Consequently, Sandra Harding (1998) announced a new historiographical revolution at the conjuncture of the sociology of science, feminist studies, and post-colonial theory.

As far as an emergent community of scholars located within science studies was concerned, philosophical and Weltanschauung approaches to the history of science made way for sociological approaches to the subject. The deconstruction of scientific rationality as a Western invention opened up the history of science to new subjects, themes, and actors—all hitherto considered marginal to the triumphalist narrative of science. Those overlooked by this vision of science included not merely cultures and civilizations outside the 'domain' of Western science, but even groups, communities, and knowledge forms peripheral to the 'cumulative growth of science' perspective. Those hitherto outside this gaze of history included artisanal classes, invisible technicians, women, practitioners of traditional medicine, blacks, coloured peoples, and many more (Douglas 1980).

A major outcome of research efforts within this widening vision of so-called science as practised within institutions of science was the recognition of the limitations of 'influence' models and 'impact' studies—in effect the unproblematic transmission of scientific ideas (Habib and Raina 1989; Raina 1999; Raj 2000). Contemporary scholarship has come to acknowledge that models drawing a one-way arrow of influence from the sites of generation of scientific and technological knowledge to society, the colonies, or industry have run their course. An osmotic metaphor captures the essence of transmission studies of science, and embodies the idea that advanced knowledge like truth flows from regions of high truth concentration (either the reigning metropolises of science within Europe, or the West in a more generic sense) to regions of low truth concentration (either the peripheries of science within Europe, or the West in a more generic sense) (Shapin 1983; Shapin and Schaffer 1985).

A number of disciplinary encounters have gone into the constitution of this revised vision of science studies. On this count it is necessary to mention the important spheres within which the issue of culture has so radically coruscated. While the constructions have forsaken the axiom of neutrality central to historical narratives of science of the first half of the 20th century, the ideology of progress, while still highly problematic, has been transmuted in terms of the science of evolution by those who would bracket 'progress'. This revision is conditioned, amongst a diversity of factors, by sensitivity to the relationship between politics and the manner in which disciplines and frameworks emerge. Contemporary studies would situate the emergence of these discourses on the terrain of a social rather than an individualist epistemology.

Returning to the history of the history of science, the Marxist externalist history of science, first explicitly enunciated in the 1930s, was to provide a tentative frame for the inauguration of the monumental efforts of Joseph Needham (1977) on *Science and Civilization in China*. One of the significant outcomes of the Needhamian project was to deny 'scientific rationality' its cultural ascription of being Western: in fact from Butterfield to Needham we have the history of science woven together by the trope of science as a cultural universal. The new map of scientific knowledge so produced, was a more ecumenical one wherein the history of sciences was embedded within the history of civilizations.

It is in this context that the question of the representation of the non-West and the histories of its knowledge forms acquires significance. While Said's *Orientalism* (1978) is a watershed in historical scholarship in this genre, it is often beset with difficulties that scholars from the non-West have indicated while being in sympathy with the political and programmatic reading that underlay Said's project. In any case, the project of comparative studies of science and civilizations inaugurated by Needham, and subsequently the Saidian deconstruction of the Occident's Orient, has not only prompted a diversity of re-readings of how the Orient was interpreted, but has also opened up the possibility of re-examining the knowledge systems of the Orient.

This re-examination today may commence from a number of standpoints. One of these would relate to how the non-West was represented in European scholarship, and how the historical narrativization of the non-West was deeply coupled with notions of Europe's historical reconstruction of its own knowledge vis-à-vis the knowledge of the non-West. Thus, in representing the scientific knowledge of, say, India, it is important to examine the place and meaning of science in contemporaneous Europe, in addition to recognizing the dialectic between the frontier of modern science and the theory of translation that underlies the reconstruction of the sciences of India. In recognizing the finer insights of Orientalist discourse it is equally important to recognize that this is not a monolithic discourse and that it contained constitutive elements that eventually undermined it.

This process is discussed in a chapter chronicling the conditions of the emergence of the social history of science in India. Any theorization of the historiography of sciences while recognizing the frame for the emergence of the history of sciences in Europe must also contend with the diversity of other ignored but constitutive elements: the evolution and constitution of Orientalism, the institutionalization of the separate sciences, and finally the relation of both history qua history and history of sciences to the proliferating nationalism of the 19th century—something that the history of science in the West becomes self-conscious of in the writings of Duhem (Cohen 1994).

Subsequently, as we move to the post-colonial period, the frame of historiography of sciences breaks up into a collage of sub-frames, wherein the post-colonial frame is the most recent and internally diverse one. The beginning of the decline of socialism, the end of scientism, which in a way commenced with the inability of the technological determinist world-view to deliver, transformed the coordinates for science studies. Significantly, the environmental crises and the subsequent deprivileging of scientific activity, as it came to be reinstalled as one of many forms of culture, enabled a different recuperation of the sciences; and for the first time we begin to witness different epistemic notions of science appearing in the scientists' account of the history of science and the social history or critical histories of science that have appeared over the last two decades.[3] Prior to the aforementioned developments in the studies of science in society, across the internal–external divide, historians shared an epistemic vision of science as transcendent and therefore as culturally universal. The essays collected in this book attempt a social theoretic investigation of the history of sciences of India, guided by an evolutionary view of knowledge, a realist constructivist theory of the production of knowledge, and a theory of history that recognizes how hitherto drafted histories are premised on a global narrative but discounts the inference that there is no ground for a global history at all. This would imply that while there exist global versions of this history, we do not have a satisfactory version thus far.

SITUATING HISTORIOGRAPHY WITHIN
THE SOCIAL THEORY OF SCIENCE

The theory of science as metatheory addresses the larger question of the opposition of theory and history. Robert Young (1990: vii) contends that in the debate on history and theory, whose genealogy dates back to the Enlightenment, the dislocating term is colonialism.

Eurocentric theory may be deconstructed by examining the links between the structures of knowledge and forms of oppression that have been with us over the last 200 years. Implicit within Hegel's philosophy and the Hegelian project of history was the 'appropriation of the other as a form of knowledge which uncannily simulates the project of nineteenth century imperialism'. This project mimics at the cognitive level the geographic and economic absorption of the non-European West (ibid.: 3).

Another standpoint from which the same theme may be addressed is the extended engagement with the representation of non-Western knowledge forms in the social sciences. This book is intended as a contribution to the social theory of the sciences. This it does from a location wherein the post-colonial situation permits the possibility of questioning the representation of non-Western knowledge in social science theory. The theory of science, unlike the philosophy of science, theorizes about science while situated within the historical discourse about science. The scientific image it contends with is that of science as practised in contradistinction to a normative ideal of science. The theory of science programmatically employs the reflective method of philosophy in order to bring reflexivity into science (Jamison 1982: 1). Furthermore, the social theory of science is a reflection upon the conditioning and determination of scientific knowledge by society around it (ibid,: ii), and it may, through the application of the same method to historical material, have a reflective bearing upon the production of historical texts.

However, the social theory of science, according to Restivo (1994: 109), dissolves the distinction between the social and the cognitive, and retains the possibility of a revolutionary critique and revision of scientific practice. Social epistemology on the other hand entertains the possibility of dialogue between the scientific community and the community of sociologists of science who apparently threaten the former. Like the social theory of science, social epistemology draws upon the resources of history and the social sciences in order to address the normative concerns of epistemologists and sociologists of science. By contextualizing these normative concerns the issues are reproblematized as concerns of evaluation and policy making (Fuller 1994: 591). At a cultural distance from the social studies of science pursued in Europe and the United States, it appears that the divide between the two is not so clear-cut. For, as already pointed out, social theorists of science are equally concerned with making 'science aware of itself', and thereby transforming the social practice of science. This task of the metatheorist of science is unrealizable if the scientist is not enrolled into the dialogue on the social conditioning of knowledge.[4] Both the social theorist of science and the social epistemologist are working towards a reflective positioning, and the real difference resides in the version of reflexivity each conscripts. The social theorist's reflexivity derives from her ability to push science towards the possibility of radical transformation. The social epistemologist's reflexivity makes sense if her normative situation resembles that of the scientists' being audited. The present work does not strictly distinguish between social theory and social epistemology, but since it is positioned as one variant of the post-colonial theory of science, it is programmatically proximate to the social theory of science.

The point of relevance here is the reciprocity between social epistemology and the enterprise of situating the historiography of sciences. Both are reflexive discourses. Social epistemology analyses reciprocities between forms of knowledge and social contexts (Elzinga 1996: 530). This perspective acknowledges that social transformations directly and indirectly generate transformations in science and vice versa. These could be at the level of scientific and social practices and values, changes in organization and, in a more mediated manner, modifications in policy agendas, and the contents of research. These transformations at several levels suggest that science and social order are co-produced. For example, the clock metaphor of the 17th century facilitated the mechanization of the world picture, as did the metaphor of struggle for Darwinian theory. The social theory of science could therefore incorporate externalities and broader institutional motives, and their embedded imagery and symbols that constitute the field of science as culture (ibid.: 537). The study of historiography is an important concern for the social theory and the sociology of scientific knowledge. The notion of co-production is relevant to such an inquiry in the social studies of science and technology. The term is taken to signify the 'inseparability of the political and intellectual' (Dennis 1997: 4). The social epistemological agenda of the idea of co-production is to unveil the implicit assumptions of the historical actors, for underlying their endeavour is 'the need to create new forms of understanding as they create new political forms' (ibid.: 4–5).

The contextualization of historiography is at the same time a study in social epistemology that incorporates externalities and institutional motives lying behind the representation and images of science purveyed from the past to the present. Crombie (1994 [vol. I]: 5) argues for a 'comparative historical anthropology of thought', addressing its development in the West on the one hand and on the other the diversity of forms of science, and the assumptions underlying models of transmission and exchange. Comparative anthropology requires that we tease out the questions posed cogently within the mental landscape of the past being studied and the responses considered satisfying by that culture. This encounter between the historian and her historical subject requires the sequential adoption of two postures. In the first stage the historian approaches the historical subject with 'explicit cultural relativism', in order to learn where and how the assumptions and motivations of the subject being studied are at variance with the historian's own: *'Vectorial treatment is of the essence of historiography'* (ibid., emphasis mine). The picture frame further informs us how the cultural ecology, comprised of the physical and biological environment, and the mental conception of existence and knowledge conditions views of nature (ibid.: 6). In the second stage the relativity of the first stage needs to be restrained by the idea of objective scientific truth and by the objective continuity of the scientific tradition. Crombie's comparative historical anthropology of thought is skewed towards fixing the coordinates and evolution of the Western scientific movement, since the objective continuity of the scientific tradition within Western historiography is a unique feature of the Western scientific movement. This betrays Crombie's own historiographic location, for his project is to reconstruct the styles of scientific thinking in the European tradition. Cultural relativism is employed as a device to sympathetically reach out to the minds of the historical subject, and the idea of objective scientific validity that is cumulative provides direction to the historical narrative (ibid.: 7). Crombie's definition that *'historiography is a dialogue between an interrogating present and an interrogated past'* (ibid.: 8), provides the conceptual link that embeds the social construction of historiography and the politics of representation within the social theory of science. The interrogated present connotes our own social epistemological location from which we examine the past, and the interrogated past alludes to the frame that we seek to contextualize. Consequently, we constantly shift between two socio-epistemic registers, and this diachronic translation is the dynamic ground of social theory.

THE TERRAIN OF POST-COLONIAL THEORY OF SCIENCE

In her book *Is Science Multicultural?*, Sandra Harding (1988) posits the convergence of post-colonial theory, feminist theory, and post-Kuhnian science studies. To begin with, Harding wishes to set up a distance from her reading of post-Kuhnian studies and that of the strong programme. This she does by introducing the notion of *co-constructivism* or *co-evolutionism* to denote the co-evolution of scientific knowledge and culture, rather than employ the concept of constructivism as is current in the literature of science studies. She wishes to surpass the antinomy of realism and constructivism. Harding argues that post-Kuhnian science studies that emerged after World War II and post-colonial science studies are approximately of the same vintage—for the period of decolonization of Asia followed towards the close of the World War II. Both these narratives posed a challenge to the standard internalist epistemologies of the sciences. As a result the trajectories (as well as perspectives) of post-colonial STS and that of the Northern co-constructivists have intersected, which is not to say that there are no points of divergence amongst them (ibid.: 4).

Post-Kuhnian studies may have initiated a major renewal in the history of sciences, but in post-colonial science and technology studies Harding sees the coruscation of a 'second historiographic revolution'. These studies bring fresh perspectives on the integrity of the European sciences, and more importantly, disclose the close interaction between the European and non-European sciences, particularly since 1492 (ibid.: 5) In addition, they have contested the script of modern science as a cultural universal, and counterposed it with a more polycentric model of the growth of several scientific traditions—albeit in some form of interaction with each other (ibid.: 6). But more than anything else the experience of post-colonial societies with development aided by the developed nations has proved the inadequacy of traditional transfer of technology models.

This inventory of challenges should be followed by a positive characterization of post-colonial theory. In the first instance, post-colonial historiography is multicultural and global in extent and range, and is at odds with the 'isolationist history' produced in the imperial West.[5] The latter histories narrated the past of non-European societies as 'largely separate and self-contained chronologies, more or less isolated from each other'. Breaking this isolation was the arrow of diffusion that commenced its journey eastward from

Europe (Harding 1998: 7) and enabled the construction of the Eurocentric history of science. The last chapter of this book discusses the globological perspective in a little more detail and elaborates upon some of the obstacles the framework would have to surmount to realize its purported objectives.

TABLE 1.1

The Distinction between Eurocentric and Post-colonial Perspectives on History, Science and Theories of Transmission

Civilizational presuppositions/ theories of knowledge	Eurocentrism	Post-colonialism
Theory of history	Isolationist	Multicultural
Theory of science	Transcendent	Contextual
Theory of transmission	Arrow of influence points eastward	Multidirectional arrows of influence constituting a network

Eurocentrism has traditionally manifested itself in many forms, and Harding specifies the various aspects of Eurocentrism. It is defined as a 'set of institutional, societal, and civilisational arrangements for distributing scarce economic, social and political resources' (ibid.: 13). At the individual level these are evidenced as *overt* and *covert Eurocentric beliefs and practices.* The vector for propagating Eurocentrism in the academic world and through pedagogy, is *institutional Eurocentrism.* This in a manner of speaking is a pedagogic and ideological programme that authorizes views about non-Europe and devalorizes claims concerning it—through disciplines, publications, and publication practices. The fourth kind of Eurocentrism she terms *societal Eurocentrism.* In this case the institutional beliefs and practices just mentioned are propagated and maintained within the larger cultural sphere, thereby reproducing this set of beliefs and practices (Harding, 1998: 13). And finally, *civilizational or philosophic Eurocentrism* refers to the prevalence of these beliefs and practices at the level of entire institutions: it is not just restricted to a subculture or a nation. In fact, this is the most problematic, for as Harding (ibid.: 14) writes: 'they structure and give meaning to such apparently seamless expanses of history, common sense, and daily life that it is hard for members of such "civilisations" even to imagine taking a position that is outside them.'

Harding recognizes that post-colonialism is a discursive space opened 'both within and after the end of formal colonialism' (ibid.: 16). Thus, it is that post-colonialism is not monolithic, and if so it could well be that there are strands in its fabric that are not multicultural either, and are the very image of the Eurocentric discourse that post-colonialism has ostensibly challenged (Elzinga 1999: 111–12). In fact, if there are epistemic points of concurrence of post-Kuhnian science studies, feminism, and post-colonialism, there are perspectives in all three discursive formations that could be embraced by post-modern epistemology as well. It is thus interesting to observe Harding steer clear of some of the leading contemporary post-colonial theorists. It could well be argued that it would be difficult to include these variants of post-colonial epistemology, since they would not serve the requirements of either strong objectivity or robust reflexivity that Harding advocates.

I do not wish to embark upon a discussion of a feasible post-colonial theory of science in this introduction, but wish to suggest that while there could be different standpoints within post-colonial theory, they are unified inasmuch as they are premised on a contextualist theory of science. But then internal fractures are manifest. Since we are agreed that we cannot speak of post-colonial theory as a unified, monolithic structure, the first feature to be noticed is that some of the standpoints lack the virtues that Harding hopes to distil from post-colonial theories. The revivalist standpoint within post-colonial discourse offers the very mirror-image of Eurocentrism in theory and history that is the subject of critique. In the final analysis Harding explores both post-colonial and feminist epistemologies to strengthen the objectivity of understandings of modern sciences and technologies, and that of other cultures (Harding 1998: 18). While we can concur with her claim that historical contextualism does not necessarily breed epistemological or judgmental relativism, would a theory of hybrid post-colonialism premised on the epistemic incommensurability of different knowledge systems conform with Harding's reconfiguration of post-colonial theory? This is an important point of departure, for she argues that her book conflicts with the epistemological relativist's position on the consequences of historical and sociological relativism; since not all proposed knowledge forms are equally good, and some may even be detrimental to the lives of the believer. Furthermore, post-colonial theorists could canvas for the claims of other knowledge systems 'rhetorically' from the epistemological standpoint of marginalized lives, and could simultaneously dismiss the protests of those marginalized lives that would embark on the programme of modernization. Opposition to neocolonial regimes could be articulated from a post-colonial standpoint that constitutes itself from a revivalist one. Thus, while acknowledging that the post-colonial terrain is a diverse and problematic one for theorizing about science, the essays that follow touch upon a few issues of concern in the historiography of the sciences of India.

THE STRUCTURE OF THE BOOK

The discourse of science displays a cyclical pattern. The connotation of the Kuhnian scientific revolution is thereby extended to the discourse about the scientific revolution. This cyclical form is compelled by the ensuing tension between scientism and romanticism that is played out in contemporary social scientific discourse. This becomes all the more evident as discourse about the sciences and the theory of science, more generally considered, came to be institutionalized between the commencement of the Enlightenment and the birth of positivism. The Orient plays an important role as a constitutive element in this discourse. In the second chapter I chart out how science and technology in India is represented between the early efforts of the Orientalists and the closing decades of the Nehruvian era. Very briefly, I mark out the archetypal fault lines in this discourse. A social epistemological reading of these concerns, frameworks, and formulations is attempted in both local and global contexts. The global context is important in our own era not merely from the point of view of post-colonial theory, but because science itself is no longer Western science, but has gradually assumed global dimensions of production. Over a period of a 100 years the key issues that come to the centre stage within romantic and scientistic accounts are diverse since they are very different political economies and circumstances. Charting out these departures and breaks in the scientism–romanticism polemic should facilitate a social epistemological understanding of the relation between the internal and external historiography of science.

The last 50 years have witnessed a departure from a particular brand of the history of science, and the figure of Joseph Needham has been particularly important in this revision. One of Needham's radical moves was to revise the theory or the history of the transmission of the sciences, without radically altering the epistemic definition of science, or what was taken to be science in different national or cultural contexts. Underlying the Needhamian framework was a certain presentism that precluded those scientific traditions or theories that did not join the streams leading to the growth of scientific knowledge. While retaining a commitment to Needham's ecumenical picture I discuss some problematic issues in the history of transmissions. These concerns can be resolved by reformulating the history of transmissions based on developments in the sociology of scientific knowledge.

For a colonial subject the inadequacy of the internal history of science in providing an answer to the Needham question was evident long before the latter acquired currency in the history of science. The social turn to the history of science towards the last decades of the 19th century inaugurated by a practising chemist and historian of science bespeaks much about the colonial context wherein colonial subjects began to rewrite their own history. Drawing upon Orientalist historiography, practising scientists and scientist historians refashioned a nationalist historiography that fed into the nationalist movement, legitimated the assimilation of modern science, and resulted in the institutionalization of science in modern India—probably more than anywhere else in the developing world. Herein lay the multifold use of history as cultural activity. The third chapter examines the relationship between P. C. Ray's project on the history of chemistry in India and his scientific researches on mercury. It is argued that this effort was part of a larger programme aimed at culturally legitimating modern science, and the two projects otherwise considered distinct mutually informed each other during three years. The chapter goes on to identify the obstacles that have impeded the recognition of the thematic unity of Ray's endeavour.

Between the two World Wars a new conception of the history of science came to be institutionalized in the West. This exhibits the culmination of the idea that it is only in the sciences and the history of sciences in particular that we have a record of the history of human intelligence. Historians of science from the West engaged with the history of the non-West by appealing to a different conception of the transmission of scientific knowledge. In the fourth chapter a dialogue between the Belgian historian of science, George Sarton, and the Sri Lankan historian of art, Ananda K. Coomaraswamy, on how the history of humankind could be written, unfolds. For Sarton the history of science epitomized the history of the development of the human mind, and this history provided a better legacy for posterity than did the history of nations and warfare. Coomaraswamy broached the problem from a different angle, informed by a deep understanding of the history of Indian art and philosophy. Coomaraswamy asserted that Sarton's historiography was beset with presentism and was founded on a positivist theory of science. Some of the methodological insights of art history or metaphysics could serve as a viable antidote to Sarton's perspective. We discuss the key themes around which this dialogue occurs, and the divergence between the two accounts.

The *Indian Journal of History of Science* was established as a journal for the history of science by a science academy in the years following independence, which were also the years for the most rapid institutionalization of science in India. The fifth chapter undertakes a bibliometric analysis of the articles appearing in the journal to infer the key areas of research in the history of science, and draws some conclusions about the sociology of the community of scientists and historians of science in India. On the one hand, while there was the underlying urgency to legitimate science by emphasizing the antiquity of science in ancient India, this search for a principle of quasi-continuity appealed to the nationalist impulse as it drew its last breath of inspiration from the freedom struggle. However, this history of science has displayed a striking conservativism and has not responded to the changing conceptions and theories of science and society. I go on to discuss a set of canonical papers that appeared in the journal, which were authored by the leading historians of science of the generation and thereby contextualize the image of science. Furthermore, the chapter explores the political dimension of the Indian state in its relationship with science that shaped representations of science in history. However, we must not forget the realm of international politics and that of global science that shaped the local discourse of science as well—and here we refer to the attempts by scientist-historians of science to repaint the big picture of science.

While it was Kuhn who gave the historiography of the scientific revolution a social turn, the historiography of the scientific revolution itself dates back a good 200 years. The 'Big Picture' of the history of science drew its sustenance from the scientific revolution that occurred in Europe in the 17th, or as others would put it, in the 18th century (Cunningham and Williams 1993). While canonizing the 'Big Picture' historiography, historians of science such as Needham and others had made it their business to ask why the scientific revolution did not take place in China, India, or other ancient civilizations that were as 'advanced' as Europe in the Middle Ages. This over-determinationist historiography inspired several historical projects both in the West and the former colonies, as they grappled with modernization and the prospect of leap-frogging into the era of modernization. As discussed in the previous chapters, the history of sciences narrated in the modernist vein dates back at least a 100 years, and not all of it was produced by Orientalists. We now turn the Needham question around to ask why a Needhamian history has not been produced in India despite the substantial logistic support for the project. Furthermore, the history of science takes off as a discipline in the years following independence, during which time the Needhamian picture had begun to prove attractive to many a scholar from the former or existing colonies. The factors accounting for the non-emergence of the Needhamian picture are both of a cognitive and sociopolitical nature, for the latter is inscribed within the cultural practices of scientific institutions.

Two sociological theories have acquired currency when discussing the globalization and institutionalization of modern science. Both in a sense are nested within each other and are premised on a sociology of emulation. The first is that of the metropolis and province, which was proposed to understand the dissemination of science within Europe and subsequently the movement of scientists and voyagers from the metropolis of science to the provinces, and the changing equations between the two sets of scientists. The second set of theories were developed by Joseph Ben-David (1984), possibly extending the centre-periphery theory of economics to the realm of scientific knowledge and the institutionalization of science. A third set of theories to be discussed in the next essay are also inspired by models of the transfer of technology and resources, and are manifest in the transmission model proposed by Basalla and others that were subsequently taken up by historians of science. The seventh chapter raises a number of points of departure from the centre-periphery model, citing the specific instance of the institutionalization of science in modern India. The essay takes up three concrete cases and subsequently elaborates upon different disciplines and how their institutionalization contravenes the centre-periphery model, of the institutionalization of the sciences in the former colonies.

In the post-1950 era, when science reaped the fruits of the optimistic vision of the 'Endless Frontier', a new world order was being instituted, which drew its inspiration from Rostowian models of growth. These models were almost co-produced with projects for the transfer of technology or the export of goods from the developed world to the developing world. Basalla's (1967) model for the 'spread of Western science' also emerged at this time, albeit clothed in the ideology of the neutrality of scientific knowledge and institutions. This model gained credibility, given its applicability to the then current policy discourse on science and technology transfer. A brief eighth chapter discusses the history of transmissions differently. Within the broader rubric of concerns relating to the Basalla model, the chapter examines the consequences of developments in the sociology of scientific knowledge and the metatheories of diffusion. It revisits the Basalla model from a contemporary perspective and proposes explanations for why the Basalla model has outlived its utility in contemporary studies in the sociology of science when applied to the study of history. What does this mean for the study of globalization of science outside the Western cultural sphere? The chapter examines the validity of the three-phase model, the sociology of diffusionism, and the politics of science and the fallout for the study of the globalization of science. In conclusion it charts out the future frameworks and trajectories for the study of science and contemporary society.

The historiography of the scientific revolution as pointed out earlier legitimated the departure from the past through a theory of modernization and the ideology of progress. Metatheoretically, the framework of modernization became a yardstick for studying the modernization of the East. However, comparative studies of different regions of the non-West have much to tell us about the different experiences with modernization in several countries of the eastern and southern hemispheres. This raises the larger question whether the Western model of modernization, the Western experience of modernization, has been the only one we have inherited or have there been multiple modernities? The historiography of modernization has been challenged by scholars working on the history of the sciences in East Asia. One of the interesting features that has come out of these studies is that the East Asian experience with modernization that commences in the 18th century is quite a contrast to the European one, for the social changes associated with modernization, that include democratization, were decoupled from those associated with the scientific revolution. Does this have anything to tell us about the metatheory of modernization as we have assimilated it in Indian scholarship?

While this study has contextualized the writing on the sciences of India, it has ignored this scholarship as it pertains to the ecological sciences and some of the more recent writing on the medical sciences of India. However, based on contemporary trends in the history of science and the renewal the discipline has undergone globally, I extrapolate the trends to suggest what these developments would mean for the future of science studies in India. How would the breakdown of the 'Big Picture', the challenge to modernization and development theory, alter the way we think about science? The answers to these questions will be further complicated by the changes science itself will be undergoing in the years to come. These changes are of a cognitive and institutional nature, and when interfoliated with the changing relationship between science and the state, it produces a mutation in the trajectory of science studies.

NOTES

1. Although Mendelsohn (1977) and van den Daele's (1977) papers could well mark the inauguration of the turn to social epistemology.
2. See also Longino (1990) and Sismondo (1996).
3. An interesting distinction has been proposed by Hess between the cultural studies of science and critical science studies. The programme of cultural studies focuses on questions of culture and power, and 'problematizes science and technology historically as part of the post modern condition' (Hess 1997: 113). On the other hand critical science studies lies at the confluence of several trends that include feminist/anti-racist studies, to the researches of the radical science group, to issues of justice and democracy.
4. Fuller's critique of the sociology of science is directed at its patent revisionism, in so much as it never rises up to its revolutionary potential. Thus, he compares the accounts of the sociologists of science as 'trying to understand the nature of Roman Catholicism by closely studying what transpires in the Vatican backroom and at High Mass' (Fuller 1995: 120–1). This would reveal little about the proclivities of the faithful. Consequently, sociologists have merely argued for a revision of the 'philosophical talk about science'.
5. The globological approach to history that Andre Gunder Frank seeks to develop, and that informs Harding's historiographic argument, is not so much multicultural as multicentric. This becomes evident if we go by his citing of Perlin: 'Whether in the ninth and tenth centuries, twelfth and thirteenth centuries, or seventeenth and eighteenth centuries, the world has always been complex in its connectedness. The continuum of medieval and early modern times has no single centre, not even a handful of particular centres conceived as a source of affecting integration. Instead its characteristic is prolific multi-centredness' (Perlin quoted in Frank 1998: 52).

2

Scientism and Romanticism

> My novel *Midnight's Children* was really born when I realized how
> much I wanted to restore the past to myself, not in faded greys of old
> family album snapshots, but whole, in Cinema Scope in glorious
> Technicolor ... imaginative truth is simultaneously honourable and
> suspect. (Rushdie 1991)

The study of the history of science and technology of India dates
back at least 150 years. Traditional accounts of the history of
artefacts, techniques, and knowledge of India are many. One of the
most acclaimed of these is Al-Biruni's *Tarikh-i-Hind*, containing one of
the earliest accounts of scientific and technical knowledge in India
and dating back to the 1st millennium AD (Sachau 1910). However,
we are referring to the historical reconstruction of knowledge and
technical practices guided by the epistemic grid of modern science,
and while embarking on this endeavour we shall examine the evolution of this discourse in India. In a 1980 review paper on the history
of science Arnold Thackray (1980) traced the disciplinary evolution
of the historiography of science over the last 100 years and identified
six broad paradigms for the history of science studies. The history of
science in the non-West was not the focus of attention.

In studies on the history of science and technology in India, the
historiographic frames are structured by a multitude of factors, such
as the nature of the interaction between traditional forms of knowledge and new knowledge. The politics of knowledge throws up
issues relating to the interplay between local cultural embodiments
and structures of imperialism. In post-colonial societies we encounter the programmes of historical reconstruction that are driven
by the desire to reclaim the 'undistorted self'. And finally, scientific
and technological knowledge are essential components of the

historiography of modernity. The difference between these frames makes it conceptually difficult to navigate between programmes whose 'historical aim' is *restoration* and those attempting the *revitalization* of scientific and technological knowledge 'put aside by Western colonial expansion and the diffusion of European science' (Polanco 1985: 309).

Further, epistemological approaches to the study of the sciences in the developing countries have had to break with the standard or Big Picture of the history of science and technology (Polanco 1985: 309). An epistemological view of the sciences in the developing countries may be grounded in a socio-economic theory of marginalization. An investigation of science, technology, and development in India has frequently been undertaken within the frames of the politics of knowledge, centre and periphery, and metropolis and province. The necessity for such a theory resides in both global and local contexts of science and politics. For the sake of convenience, four phases in the evolution of this discourse are proposed. The discussion that follows will naturally highlight the differences between these phases and the contradictions implicit within each of them as well. These four phases are:

- British and French Orientalist studies on the sciences of India;[1]
- pre-independence nationalist studies;
- the phase of post-colonial reconstruction or the golden age of scientism; and
- the postpositivist phase.

Each of these phases is naturally stamped by the raging socio-political concerns of the time, and in each of them we witness variants of the contest between the scientific, romantic, and social realist images of science (Elzinga 1984). In the following discussion, homologies with historical discourse from other cultures and civilizations will become evident.[2] Further, there will be homologies as we transit from phase to phase, but the resolution or closure of controversies in each phase will differ.

THE AGE OF ENCHANTMENT

In the 18th century, science was undergoing a rapid disciplinary institutionalization (Gillispie 1980; van den Daele 1977) in Europe. Modern histories were produced at this time that served several objectives, some internal to the disciplines and others relating to the legitimization of modern science in contrast to classical learning (Laudan 1993). Three discourses were important for the constitution, to take an important discipline, of the history of mathematics and astronomy. The first of these was a proto-positivist theory of science, developed by Condillac and those around him that later culminated in the work of Comte. The second was the history of these disciplines in Europe and the third, the history of mathematics of the non-West (Raina 1998). Christian missionaries, British voyagers and administrators, French voyagers and savants, among others, wrote the history of sciences of India, framed by the historiography of modernity in the 18th century. These early accounts comprise histories of the indigenous systems of natural knowledge as well as the reservoir of techniques available in these societies.[3] Hence, from the *Lettres Edifiantes et Curieuses* (1810), five of the 16 volumes consisting essentially of letters of French Jesuits visiting India,[4] to the fictive discourse of a Voltaire that sought to signify Europe differently, India was presented as an *etre-ailleurs*, and simultaneously signified, as Murr (1986: 15-21) points out, in three different senses:

- as *radical alterity:* India signified as an 'elsewhere';
- as *similitude:* India signified as an other here; and
- as *exemplar:* India signified as Utopia.

The writings of the French Jesuits in India are marked by ambivalence towards the renaissance and scientific thought sweeping Europe. Their writings on Indian customs, theology, and jurisprudence are simultaneously stamped by fascination, enchantment, and criticism of a new world with roots in antiquity. As part-time scientific explorers for the Académie des Sciences, their reports of the historical astronomy of China and India fine-tuned the consolidation of celestial mechanics (Raina 1999). The reports sent back to France were to form the basis for the histories of science produced during the Enlightenment (Murr 1983; Raina 1999). While the French savants were long indebted to the Jesuit corpus on India, their optic was ostensibly different, and gradually this debt to the Jesuits was written off in the triumphal celebration of Enlightenment. Between 1760 and 1850 a number of histories of mathematics and astronomy of India were discussed either within the context of the world history of mathematics or the specific Indian case. Other than the limitation in sources, these histories were embedded in a complex cosmology of 18th-century Europe, where mathematics had acquired pride of place. With the rise of the modern nation state, nationalist concerns soon permeated this discourse and oriented its direction by sensitizing historians to concerns of priority and prestige. This tradition of the history of science entered into a polemical debate with the British Indologists. This divide between distinct national interpretive traditions was also precipitated over questions of scientific authority (Raina 1999).

The British Orientalist tradition, implicated in a policy of imperial rule, conceived the Enlightenment as a double-edged sword. James Mill attacked Voltaire and the *philosophes* for their sentimental attachment to the Orient, while concurrently being uncomfortable with 'Macaulay's extreme chauvinism' (Kopf 1969: 237). Colebroke's works on the history of astronomy and mathematics in India pre-dated the Anglicist–Orientalist controversy. In fact, he was instrumental in publishing one of the first translations of the mathematical works of Brahmagupta and Bhaskaracharya in English. However problematic the translation, this was to have a remarkable influence on a whole generation of mathematicians (Augustus de Morgan and George Boole being notable examples) as well as 19th-century historians of mathematics (Colebroke 1873).

An important item on the agenda of the British Orientalists studying the natural knowledge systems of India during the early years was that of exploring the possibility of grafting modern scientific knowledge on to a Sanskritic base. Long before the Orientalist–Anglicist controversy broke out, and well before the institution of the Macaulayan diktat, it was considered a 'visionary absurdity' among the Orientalists to embark on any pedagogic endeavour in English (Kopf 1969: 242). The Orientalists and Anglicists shared the same 'universalist faith' in the substantive transformation of the non-Europeans, but the point of departure was how this transformation was to be effected.[5] For the former, this was possible through a syncretistic effort aspiring to transform tradition itself; in the latter scheme a programme of erasure was to be followed by one requiring 'assimilation on the British cultural pattern' (Kopf 1969: 246).

Both French and British narratives on the sciences of India were constituted in terms of a multiply-oriented grid. On the one hand they were urged on by *origin and influence questions*. There were several strains within Orientalism concerned with the presentation of India as an object of knowledge: the British, French, and German Indological traditions were the dominant amongst them. Along with the discovery of affinities between Sanskrit and the European languages, there was also the ongoing philological project of discovering the origins or childhood of Europe in India (Prakash 1990: 385), but as much in the realm of the history of the sciences of India. Depending on how the questions were framed, varied spheres of investigation and approaches to addressing related problems emerged. The Eurocentrically oriented would seek out Greek influences in Indian astronomy or medicine—rather than frame the problematic within the context of exchange. For the more romantically inclined French Indologists, the India of antiquity was exemplar—a Utopia, from where could be culled gems of scientific wisdom that had escaped Europe but could enrich it all the same.[6]

SCIENCE AND NATIONALISM

Orientalist scholarship on the sciences did not come to an end with the commencement of the nationalist phase; in fact both co-existed and informed each other at this time, persuasively and dissuasively. However, a whole new range of questions was posed during this phase, responses to which necessitated the adoption of distinct historiographical frames. If the colonialist writing on India was essentialist, the burgeoning national consciousness among Indians was inscribed along economist lines (Pandey 1992: 166). Economism served as a modality for explaining Indian underdevelopment, in dis-identifying with the essentialist gaze of the Orientalist as much as with that of colonial historiography. Prakash has pointed out the commonalities between the Orientalist and nationalist writing. Other than the transformation of an inert, passive object of knowledge into an active, sovereign one, nationalist history brought regional history into focus, challenged the assumption that 'all that was valuable in world civilization originated in Greece', and also challenged the interpretations of the chauvinists among them. Nehru for one elaborated upon the cultural and historical diversity of India so that 'India could join the world historical march towards modernity' (Prakash 1990: 389).

In the late 19th century the crisis of 'legitimate authority' occasioned by the expansion of the colonial state created the need for narrativizing the past, wherein 'history itself became the site for struggle between contesting parties' (Prakash 1992: 151–2). Nationalist discourses at this time wore two faces, the more dominant among which derived its modernist idiom from 'the metropolitan culture of the colonizers'. From the ranks of these Western-educated secular publicists emerged the cadre that would lead the cultural

counter-offensive (ibid.: 207). This offensive, in a decidedly dialectical fashion, was directed against what Said (1978: 206) has called the
'binary typologies' produced by Orientalism, these being the notions
of the advanced and backward races, cultures, and societies. Consequently, the predicament of 'racist colonial science' was its inability
to escape the 'liminality produced in its own performance. As the
colonizer staged the colonized as man, he disavowed the racist
polarity ... that was the enabling frame of his discourse' (Prakash
1992: 160).

The seeds of a modern indigenous scientific tradition were first
sown in Bengal, where in the late 19th century a consciousness and
deliberation on the nature of history broke out 'everywhere'. As
Kaviraj (1988a: 1) puts it: 'history is the privileged discourse of nineteenth century Bengal', a discourse that was imbued with a dual
significance, being both the name for 'hope and despair'. For the
Bengali *bhadralok* intellectual, the European rationalist conception of
history was that of a 'reliable account of a people's past', a conception
that was considered worthy of assimilation and possibly emulation
as well; but they simultaneously endeavoured to transcend it in
giving this history an 'imaginative unity' (Kaviraj 1988a: 4). These
two aspects of history were themselves the product of a 'colonial
asymmetry' that predicated that the duplication of the model of
European history was impossible.

A battle of ideas was staged in this 'theatre of truth and imagination' where, to reiterate a Hobsbawmian point, the native interlocutors sought out ways of reformulating the idea of progress (Hobsbawm
1988: 15; Raina and Habib 1993). In the attempt to surmount colonial
reconstruction, the colonized would reckon with the constructedness
of the notion of history, and this idea would find expression in their
constructions. Naturally, this entailed writing a presentist history.
There are two related modalities of interrogating the past. The insertion of new theories or knowledge in the sciences is often legitimated
by revealing its connection with the past (Thackray 1980). Further
interpretations of the present appeal to the past in order to settle
whether the 'past is really past' or 'whether it continues, albeit in different forms, perhaps' (Said 1994: 9). Between 1896 and 1912, Indians
wrote two very important histories of the science of India in English
(Ray 1902; Ray 1907; Seal 1915). These two works were canonized
within the nationalist historiography of science, though Ray's *History*
always occupied a rather precarious position amongst the nationalists. But even within the nationalist frame the structure of these

works and the representations of science current among the *bhadralok* community are particularly interesting, because we have here the prefiguration of a genealogy to the scientism of the 1950s and 1960s. This is of interest since Comtean positivism was to find a voice in the educated enclaves of the Bengali community by the 1860s through the efforts of Harish Chandra Mookerjee and the British positivists like Congreve (Forbes 1975). However, as is the case with the translation of theories and knowledge systems in divergent cultural contexts, during the first few decades of positivism in India, Bengali concern focused a great deal more upon Comtean religion than on the elaboration of the structure and growth of scientific knowledge— its attainment of scientificity. This apparent oddity had to do with the fact that the inaugurators of the project of modernity in India, amongst whom is counted Raja Rammohun Roy, had already recovered a 'rational theology' from the traditions of Hinduism (Sarkar 1975a), which rendered the new knowledge commensurate with traditional wisdom. This process of legitimating new knowledge forms alerted the epigones of tradition, who may have rejected the demand for assimilation.

These influences are deeply indicated in Bengali literature, popular writing on science and culture, as well as in the more scholarly disquisitions of Indian interlocutors seeking a plane of 'cultural redefinition' (Metcalf 1986). The critical dialogue initiated by these Bengali cultural amphibians, to borrow a term from Said (1990), enabled the neutralization of the cultural and political meanings imparted to the term modern science. This further legitimated the casting of modern science as morally worthwhile and economically beneficial. Recapitulating the two epistemic positions vis-à-vis traditional knowledge, the imperial or Macaulayan position saw all the knowledge of the East as not fit enough for British schoolgirls. The dismissal of traditional knowledge as lower-order knowledge rationalized the imperial agenda of erasure and civilizing through re-education.

The revitalist project on the other hand attempted to recover and reconfigure all that was worthwhile and salvageable from the indigenous traditions of knowledge, and in the process sought to establish a series of *transit points* into the realm of post-Galilean science.[7] The activity of the promoters of science, cultural amphibians as they were, was reflected in their choice of research problems, and in their attempt to derive or invent cognitive homologies bridging the two traditions. In this pre-Sartonian era historians of science were practising scientists, such as P. C. Ray, who in treading into the past not merely triggered off processes of trans-cultural conceptual bridge-building, but were stirring up tradition in search of nuggets of neglected knowledge or possibly research problems.[8] This is evident in Raman's early work on musical instruments or in Ray's notable work on the chemistry of mercurous and mercuric compounds, to be discussed in the next chapter.

In the present chapter I shall briefly discuss the representation of science in these two canonical works. It is noteworthy that these works on the history of the sciences were published just when the freedom struggle in Bengal (Raina and Habib 1995) was entering its most militant phase.[9] The exemplar of the sciences was the so-called positive or exact sciences. Consequently, the historical enterprise required recovering those elements of traditional knowledge that were commensurate with the epistemological ideal of science as positive or exact knowledge. Both these terms imbued science with its 19th-century self-image of generating reliable knowledge.

The ideal of science as a cultural universal or transcendent entity enabled the Indian historian to recover or shelve elements from the Indian tradition, and served the secondary objective of establishing intellectual parity with the West. In the age of nationalism science symbolically represented the progress achieved by a nation, and the history of science was the field for symbolic warfare as well. With the decline of the scientistic ideal in the 1970s the historiography of science as a cultural universal lost its shine, and the methodological imperative that had guided the first generation of historians was abandoned by most cultural theorists of science. For example, a matter of debate between those wishing to promote herbal medicine and the practice of ayurveda was that the efficacy of drugs prescribed in these systems of medicine cannot be validated through double-blind tests. Some proponents of the allopathic system have suggested that surely these practices could be submitted to procedures of validation that were relevant within the traditional systems.

B.N. Seal's (1915: 241–91) *The Positive Sciences of the Ancient Hindus* is best known today for its chapter on the scientific method and its reflection in the systems of Indian philosophy (*darsanas*). For Seal, a methodological review was absolutely essential if an understanding of the positive sciences was desired. Seal deduced that as far as the sciences of ancient India were concerned, the whole 'movement was genuinely and positively scientific, though arrested at an early stage'. Seal chose the schools of Buddhist and Nyaya logic for his exposition, and re-presented it in the idiom of inductivist philosophy of science. Large sections of the book deal with criteria or tests of truth, perception, observation, experiments, fallacies of observation, the doctrine of inference, ascertaining causality, synchronicity of cause and effect, and relation of time to cause and effect. This reconstruction has been vulnerable to the criticism that it is predisposed to logical and philosophical speculation (Clark 1937: 335). However, the modern scientific tradition was seen as largely empirical, and an attempt was made to recover cognitive homologues of empirical and utilitarian knowledge from the Indian traditions. Put another way, this project of revitalization was organized in terms of a method-ological recasting of apparently disparate knowledge systems in order to render them commensurable with the larger programme of modern science. My concern is not with the authenticity or accuracy of this history, but in recognizing that it was predicated on a principle of, shall we say, *quasi-continuity*, wherein it would be possible to subsume different scientific traditions, located within distinct epistemologies, within the exemplar of a unified science. While Seal's focus was on the inductive sciences, Ray on the other hand, as a working chemist and certainly the founder of the tradition of modern chemistry in India, was to provide the experimentalist's perspective. He sought through a careful interpretation of textual material to posit the basis of an experimental science like chemistry in India. Ray's work for the first time provided a grid for the interpretation of sciences of ancient India. Two high points thus appear on the horizon of scientific discourses in India.

The first is a universalism that characterizes the history of science, and for partly politically obvious reasons is devoid of the fundamental Eurocentrism of the age. Second, and not paradoxically, an engagement was posited between the growth of scientific knowledge and that of the emergent programme of nationalism. Science had come to be considered the yardstick, at least among members of the scientific community, against which the progress of a nation was to be measured. This, logically, led Ray to pose the central question that repeatedly resurfaces again and again in science, technology, society (STS) studies in developing country contexts, namely, what were the conditions that prevented the realization of the scientific revolution in India? (The question could also be posed in the Chinese, Arab, and myriad other contexts.) The response to the question could not have been found in the differing logic of the sciences of distinct culture areas. Internalism was found inadequate, and Ray had to locate the causes *of the decline of the scientific spirit* in ancient India, outside the proximate realm of scientific practice itself (Chattopadhyaya 1986: 14). This move to the external history of science simultaneously accomplished two tasks, although it would be stretching the point too far to insist that there was an element of premeditation about it. First, it paved the way for the departure from colonial historiography. Second, while drawing upon Orientalist scholarship, it also broke with a fundamental postulate of Orientalism that premised an ontological and epistemological divide between the Orient and the Occident.[10] In terms of historical scholarship, Ray had been deeply influenced by the mid-19th century histories of chemistry. These narrativized not merely Berthelot's millennium, but also a new internationalism situated in a universalist faith. In conclusion, it may be pointed out that the process of historical reconstruction that commenced during the phase of nationalist struggle was simultaneously imbued with positivist faith, a Baconian optimism in science, but had awakened rapidly to the limitations in the internalist account of science.

THE GOLDEN AGE OF SCIENTISM

The notable debates during the most intensive years of the nationalist struggle and preceding the recognition of Indian independence were carried out in the pages of the Calcutta review, *Science and Culture*, and were informed by the deliberation of the National Planning Committee. The primary concern of these years related to the nature of scientific institutions to be established in the country, the research agenda for the future, the gradual initiation of the debate on science and planning, the linkages of the scientific R&D system with industry, and the establishment of a national innovation system.[11] The ground for the emergence of a scientific and technological R&D enterprise in independent India appeared to have been prepared during these years. With the benefit of hindsight it may be suggested that these deliberations were guided by consideration of national sovereignty, and two research imperatives served as axes along which subsequent institutionalization would occur—the industrial and nuclear research imperatives (Raina and Jain 1997). The period 1920–40 is very crucial for it can certainly be considered the period when science policy and planning discourses in India acquired a concrete form. However, the constitution of this discourse is not the subject of this chapter, suffice it to point out that in its origin this discourse was imbued with the scientism inherited from the previous phase.

Yet in epistemological terms the two phases stand out in contrast to each other. In the former phase a science-based ideological social critique did not privilege scientific knowledge over other forms of knowledge or practice. The physical realm was addressable in terms of the method(s) of science(s), while there was an inner moral realm rooted in a rational theology—this partitioning of the two worlds bears a striking Kantian affinity (Raina and Habib 1996). In the latter phase there was little latitude for methodological or epistemological pluralism even within the sciences. This phase of scientism was to continue into the 1970s. Bernalism was to be a noteworthy influence during this period. The immediacy of politics and the ideological battle in the political realm were to fortify the scientistic strain that marked the progressive politics of social transformation.

The early years of the Nehruvian era were stamped by this scientistic vision, and the elites of Indian science were able to enlist the political elite in furthering investment in scientific and industrial research. The development paradigm of the 1950s was constituted of three elements: modern science, planning, and industrialization. Centrality was accorded to scientific and technological knowledge, and these were to be deployed for the task of social transformation. To reiterate a point made above, the influence of the Bernalists was fairly strong in these years. Between 1948 and 1953 the inaugural functions of the new industrial research laboratories were graced by the iconic figures of Blackett, Haldane, Bernal, and Frederic-Joliot Curie. In fact, Patrick Blackett was amongst the few foreigners between 1947 and 1972 to have played a significant role in initiating discussion of the 'role of scientific research in military developments', and actively promoted 'the careers and conditions of scientists attached to defence research' till 1964 (Anderson 1999a 1999b)

In the cultural sphere one of the first tasks to be undertaken by a post-colonial regime was that of rewriting its history. Generalizing across the spectrum, two principal historiographic programmes emerged, with a number of positions between the poles. One was ultra-nationalistic and an inverted image of the Eurocentric account, the two sharing almost similar norms of evaluation. The other drew inspiration from the anti-imperialist struggle and was situated within Marxism. Irfan Habib (2000) writes: 'With Independence, new questions within the stream of historiography were generated. As the direct compulsions of debate within British Imperialism receded, there developed a greater readiness to study the factors of change and stagnation in our past and to identify various internal economic, social and ideological contradictions' (Habib 2000). This divide between the ultra-nationalists and Marxist discourse permeated the narrower context of the history of science writing. The ground between the two was occupied by liberal historiography, rooted in the image of a value-free neutral science and technology. It appears that scientism spread across the spectrum of positions, a phenomenon that was quite characteristic of the 1950s. Consequently, science was one of the transcendent cultural universals of the age—and the faith in it was not disputed in the 1950s. The more parochial genre of history of science, while upholding the universality of the science postulate, sought to invert the amnesia of the Eurocentric history of science and technology by reversing the purported flow of ideas. This project of redressal engendered a series of priority disputes relating to discoveries that were venerated in the history of science and technology. On the other hand the *Indian Journal of History of Science*, to begin with, departed from this chauvinist framework, publishing articles and unearthing a wealth of source material providing for an internalist textual reading of the history of science. The scientists' history of science is discussed in a subsequent chapter.

Marxist historians of science, who in turn trace their origins to the 1931 History of Science Congress, inspired the social history of science in India as elsewhere. Post-colonial India was and continues to remain a field marked by polemical debates on 'versions of history'. The Marxist history of science ran counter to both Eurocentric history as well as internal colonization within the country. Chattopadhyaya (1986) was to write:

The factors that inhibited the development of modern science in Indian history are inclusive of those that are still creating the zeal for casteism and communalism ... we meet the same monster from whom inspiration is still being drawn, often surreptitious enough though also overtly. ... A study of science in Indian history is more than a mere academic exercise. It is also linked up with the question of our survival.

Second, social history ran headlong into the construction of India in Western social theory. The problematic dichotomies were those of tradition–modernity, spirituality–rationality, spiritual civilization-materialist civilization. The 19th-century Occidental representation of India as a spiritual civilization and as a counterpoint to Western materialism needed reassessment.

The contrastive difference between European and Indian

civilization had to do with that between materialist and spiritual civilization. In the former rationality permeated 'the whole of capitalist culture', while for Max Weber Indian civilization rather than being driven by the rational drive for capital accumulation was creating 'irrational capital accumulation for magicians, mystagogues and ritually oriented strata' (Thapar 1993: 50–1). The new project was to unveil the scientific and technical core of the 'spiritual civilization', if such there was. The disclosure of these traditions required responding to some fundamental questions, responses around which would be organized subsequent texts on the history and philosophy of science. The first question that bothered historians of science, in the contemporaneous context, Elzinga has framed in a broad philosophical sense. This he calls the central question in non-Western contexts: 'To what extent was the appearance of modern science in other areas of the world a consequence of cultural factors, and to what extent was it a concomitant of the non-development of specific socio-economic conditions associated with the privatised profit-motive of capitalism?' (Elzinga 1981: 1).

The internal history of science did not provide adequate answers to these questions for it never addressed science as a social or cultural activity. The project of Debiprasad Chattopadhyaya was to identify the obstacles in the path of the flowering of the scientific tradition in India. Informed in the first instance by the monumental history of Indian philosophical thought authored by Surendranath Dasgupta (1980), Chattopadhyaya (1959, 1976) went on to recast this history of thought as the battle of ideas between contending classes. The Marxist E.M.S. Namboodripad (1993), in his obituary to Chattopadhyaya, was to point out that among the many findings of the latter's work was that the ideological struggle between materialism and spiritualism in India was manifest in the social struggle between the large majority of lower castes and the upper castes.

Commencing with the traces of materialist thought in Indian antiquity, Chattopadhyaya explained the eclipse of the scientific tradition in terms of Marxist interest theory: class conflict in India or the battle of ideas between the hieratic and landowning classes and the artisanal/labouring communities. This was reflected in the suppression of materialist thought of the latter by the former. The conjunction of theoretical and practical knowledge or the crafts was seen to be a factor contributing to the scientific revolution. The absence of such a conjuncture in the Indian context could be called on to explain the non-emergence of modern science in India. This Baconian conception was to come down from the 19th-century social critic of a caste-based social order in India.[12]

Chattopadhyaya's later work was inspired by the Needhamian conception of the history of science and technology, and as a project gave concrete form to science as a socially transformist force, challenging obscurantism from age to age. In fact, a feature shared by the entire Cambridge left, and Bernal in particular, was that dogma and superstition posed obstacles in the way of scientific curiosity; and the main agency of dogma and superstitions was the institution of religion (Ravetz 1992: 157). However, underlying this idea was the attempt to portray science as an alternate form of high culture. Chattopadhyaya's early work was based on the interpretation of philosophical tracts. The latter work investigates 'artefacts' as well as other 'material' evidence, and in the process raises a number of broader questions on the history of science in ancient India.

Other than one book on the social context of the emergence and decline of theories of medicine and the practices of surgery in India, Chattopadhyaya's work (1978, 1979, 1982) contextualizes within a Marxist framework the methodology and epistemology of science as articulated within the logical and philosophical traditions, and ascribes the agonistic contest between schools to the divergent interests of social classes.[13] Inherent in the programme was a deeply polemical question, that of educing historical evidence that could set the terms for formulating a well-grounded critique of Indian caste-based society. The non-occurrence of the scientific revolution in India was causally locatable within the schism between theoretical and practical knowledge, socially embedded in the rupture between the upper castes and the artisanal and agricultural castes.

Within the same tradition, historians of technology and economic historians were researching the other central question: in the light of the pre-colonial evolution of the industrial base in India, could conditions have been ripe for an industrial revolution? That these questions were posed within separate disciplinary frameworks, ignoring the mediation between them, tells us that in the 1950s the Indian academic community saw the enterprises as distinct and cordoned off from each other. Returning to the condition of India's production base at the time of the onset of colonialism, it has now been well documented by economic historians that the profitability of Lancashire and Manchester in the 19th century also had to do with the systematic demolition of India's textile base (Baber 1996, Macleod and Kumar 1995).[14]

Thapar (1993b) points out that the history of pre-colonial India had witnessed three major changes of paradigm. The first commences with James Mill, who saw Indian civilization evolving out of three civilizations, Hindu, Muslim, and British (a classification that Rahman suggests that Sarton was to subscribe to later), and justified the British conquest of India. The second was that inaugurated by Vincent Smith justifying colonial rule, which the nationalist historians were to reverse. Finally, the Marxist shift of the 1950s was ushered in by Kosambi. D.D. Kosambi initiated research into the history of technology. In fact Kosambi is the founder of the Marxist paradigm of history writing in India—and not merely that of the history of technology. The new tradition of historical studies privileged artefact and practice over the text. As Kosambi wrote: 'If ... it is more important to know whether a given people had the plough or not than to know the name of their king, then India has a history. *History is the presentation in chronological order of successive changes in the relations and means of production* (emphasis added).[15] This historiographic orientation was to provide the frame for the disclosure of the social embodiment of technical development: 'A new stage of production manifests itself in formal change of some sort; when the production is primitive, the change is often religious. The new form, if it does increase production, is acclaimed and becomes set .... If the superstructure cannot be adjusted during growth, then there is eventual conflict' (Kosambi 1985: 12).

Despite the fact that Kosambi refrains from deploying economic determinist arguments, weak economic determinism persists as a modality of explanation.[16] In a sense the rationality of science is privileged over other kinds of rationality, for in each age the homologue of this rationality is considered the principal dynamic of change. Interestingly, this conception of science resonated in the scientific community during the post-World War II years, when science came to occupy a central place in contemporary culture. As Fuller has pointed out, the picture of technologically neutral and value-free science was purveyed during the Cold War years, which in turn legitimated science as ideologically transcendent. Politicians seeking investment in science were not seen as promoting an ideology other than that of development and modernization. In the Indian context this is manifest in the symbiotic relationship between an all-embracing Nehruvian programme and the Bernalists in India, who had a significant role to play in shaping the future of scientific institutions and science in the country.

In addition to ideologically legitimating science during the boom years of science in the country, this historical enterprise simultaneously challenged regional or continental essentialisms, while creating a space for a science-based transformist politics. These were also the boom years in heavy industries, with the Indian state investing in the establishment of public sector industries. In the realm of international politics nations had been ideologically polarized by the terms set by the Cold War. Within the Nehruvian framework India had supposedly embarked on the path of socialist development, and Nehru's charismatic personality as a statesman of international stature was to provide India with the leadership of the Non-aligned Movement. The foundations of the new society had to be fortified by the ideology of science—this ideology was to be Bernalist-Nehruvian.

At about the same time the principal focus of historians (and not historians of science or technology) was agrarian relations and the impact of colonialism on Indian society. These studies threw up a series of questions that were organically tied up with the themes of industrial revolution and colonialism. If the cardinal question posed by historians of science was to decipher the conditions that served as obstacles to the realization of the scientific revolution in India, the others were asking: 'Why India failed to industrialize (and develop a capitalist economy) either before or after the British conquest' (Habib 1971: 1). As a research programme, it could have been problematized from a host of perspectives, anti-imperialist or otherwise, but for economic historians the relevant problematic was reducible to the 'potentialities of development in the Indian economy prior to the British conquests'.

Understanding the social contexts within which technology was grounded was problematically a rich area, as a consequence of which the technical or the technological history of technology or techniques remained largely neglected. However, there is one work of this externalist genre, Qaiser's (1982). The cardinal issue continued to remain unresolved. Economists and historians in the Marxist mould, while cognizant of the limitations of the notion of the 'Asiatic mode of production' nevertheless located their problematic within a more 'universalist' framework.[17] However, the framework was not as uniform as it appeared. In the act of economically and culturally cognizing non-European societies, a diversity of problems and issues was to enforce conceptual modifications.

The plasticity of the interpretive mould depended upon the models of historical explanation deployed. Working within a universal script meant being guided by a particular model of interpretation. This virtually meant canonizing an exemplar of the scientific revolution, and identifying the factors that were present at the time of its realization. In novel contexts, the absence of the scientific revolution was explained either by the deficiency of one or many causally related factors from the canonized narrative. It is only with the emergence of a historical anthropology that a departure from these paradigmatic narratives in the history of science came to be produced—alternate historiographies were to be concretized with the notion of alternate sciences.

THE CULTURAL HOME OF SCIENCE

The rate of growth and multiplication of scientific institutions reached saturation in India towards the latter half of the 1960s. But the fillip given to scientific and industrial development in the previous decades, in consonance with centralized planning of the economy, created the demand for institutionalizing science policy studies. With the sole exception of the Jawaharlal Nehru University, New Delhi, which created a department of science policy studies, the university system failed to respond to this demand. This possibly had to do with the marginal place of science, as well as with the perception of science in India as peripheral or derivative. Abdur Rahman, belonging to the industrial research system, nevertheless pressed on for an institute dedicated to the science of science studies—a commitment that saw fruition in the early 1980s with the founding of the National Institute of Science Technology and Development Studies. The institute's charter ran a gamut of concerns spanning science policy to other domains of science of science studies, such as the history and philosophy of science, and the studies that subsequently appeared revealed a range of ideological positions and disciplinary backgrounds that reflected post-colonial concerns.

Rahman's (1982a, 1984, 1987) writings on the history of science in India draw out the contours of a comprehensive research programme. In fact, this history is prescriptive to the extent that it sets out how the history of science in post-colonial India is to be rewritten. The research programme problematized the history of science and technology studies in the Indian context and the inherent limitations with available historiography. Rahman pointed out that Sarton, when speaking of the sciences in the non-West and India in particular, inherited a fundamentally imperialist partitioning of Indian history that was hardly sensitive to the seamless texture of Indian culture; so much so that Sarton provided a sectarian account of the history of science in India, for example, even his periodization—the Vedic period, the period of Muslim rule, and the subsequent period of British rule—was sectarian. The periods were thus characterized not in terms of categories intrinsic to the growth of scientific ideas, but in terms of the ethnicity or religious identities of the ruling elites.

The programme charted out by Rahman was Needhamian in character; with its faith in the ecumenical character of science, it emphasized that modern science historically speaking was 'a product of many civilizations and the common property of all mankind'. In other words it was to be an analogue to the colossal archaeology into China's past. China and the Arab world had produced their encyclopaedias of science and technology, but where were the Indian scholars who would help to complete the tale? In this plea resided a sense of balance as imperative, a need to damp the extreme and erratic sway of the pendulum—neither Eurocentrism nor Eastern romanticism. The universalist frame of science acquired consistency through the concrete cognisance of the historicity of scientific knowledge in distinct cultural contexts.

Although Rahman himself did not give substance to this vision, he created the institutional space within which the vision could acquire detail. By the time he could provide for this space, the 1950s and 1960s vision of Indian science was undergoing a radical transformation. By the early 1970s the fortresses of centralized development came crashing down. The Bernalists in the scientific community were slowly going into eclipse; consequent to the romanticism of the 1960s, both anti-science and Eastern mysticism proved to be the counterpoints to the (alleged) ideological neutrality of science. The radical relativism of anthropologists and sociologists was, within academic circles, to contest the epistemic privileging of scientific knowledge and method—science could no longer rule supreme in the name of truth. As a result, the terrain of ideological combat had come to be defined afresh. In Rahman's scheme of things radical incommensurability and relativism did not hold much water, for true to Needhamian ecumenism, the tapestry of scientific knowledge was woven through moving centres of learning. There is, however, in his latest writings some ambiguity concerning the contention that sciences grounded on distinctly Indian epistemic premises could function with more benevolent consequences in India than does the regnant paradigm of science in non-European contexts.

The post-colonial reconstruction of the history of culture, science, or ideas has been a project in the recovery of the self in the former colonies, of overcoming marginalized knowledge forms, practices, and traditions. At one level the project of the recovery of the self indirectly impacted upon a project more germane to the sciences—that of enriching science and techniques (Visvanathan 1997). The project was inspired by the realization that not all the knowledge that had been swept away under the colonial mat was worthless. The politics of knowledge came to be constituted as a legitimate domain of investigation. The processes of colonial appropriation of traditional knowledge within the ongoing programme of modern science was an important research theme. A related theme focused upon the processes of the marginalization of traditional knowledge within the framework of colonial policy. Closer to contemporaneity, policy oriented research addressed the inauguration of the project of modern science in India. This threw light upon the current legacy of scientific institutions and programmes, and had a bearing on science and technology policy. It may be suggested that J.P.S. Uberoi (1978, 1984) may be considered the initiator of the sociological tradition within science studies in India. His writing makes a substantial plea to Indian scholars to subject all theories, modern and ancient, Eastern and Western, to fresh scrutiny and judgement. Modern science in this view had set out on the wrong path (Uberoi 1978: 15), and this critique of modernity and modern science prompts a search for other modernities and alternate sciences.

The project of recovery of the pre-colonial self had its repercussions in the history of science. In Rahman's framework this pursuit was still grounded in progressive politics, seeking as it were to posit the multi-dimensionality of scientifically constituted knowledge. Although never stated in the language of the sociology of knowledge, the underlying assumption was that knowledge was formed and shaped within particular historical and social contexts. The crisis in development of the 1970s spawned anti-science movements and a search for alternatives. The project of modernity came up for questioning. The new programmes posited different world-views. The earlier romantic argument was now reformulated in terms of a critique of progress and of Western hegemony. Internationally, the socialist dream tottered, while the scheme of large-scale centralized industrialization generated its own range of both social and economic problems within Western society. Occidental society turned eastward in search of an alternate vision of nirvana, and the Orient turned inward. This is why, as Restivo (1983) writes, the parallelism drawn between the insights of modern physics and eastern mysticism could be interpreted as an attempt to paint science with a human face.

One response to this crisis was reflected in the search for an epsitemologically alternate science. The turn to the autochthonous, it was felt, would ensure that these other sciences would be compatible with the supposedly far more benevolent telos of Indian civilization.[18] Two factors appear to have shaped this response. First, the crisis in development surfaced as much in the institutions of science and technology; Indian science was now perceived as living a dependent existence. The halcyon days of Indian science and technology of the pre-independence and Nehruvian eras had since passed—the world of science had become simultaneously populated and alienated. The dependent existence of Indian science was evident in the aping of Western research priorities and programmes in the sciences (Reddy 1978a, 1978b). Second, indicative of this alienation was the recognition of the lack of connection between science and industry, and science and society. For science to be rendered relevant to social demand, it had to be sensitized to the local cultures and contexts.

The social studies of science as pursued in the West could be framed under three broad headings: academic liberal studies, technocratic and responsibilist, and criticist (Elzinga and Jamison 1981: 2–3). The social studies of science in India grew out of an engagement with politics and from social movements, wherein most researchers ended up treading a tightrope strung between research and activism. Thus, by the 1970s and 1980s, policy makers were still committed to the technocratic image of science in social transformation, while scientists associated with social movements were opposed to the positivist conception of science. The criticist response was variegated, incorporating romanticist, populist, and strains of a dying positivism. But the internal fractures within this discourse were situated along the dichotomy wherein the East was conceived as experiential, aesthetic, intuitive and the West as experimental, rational, theoretic (ibid.: 1981: 9).

As referred to earlier, both scientism and romanticism share common features. The 'immaculate conception' of science (Visvanathan 1997) as a value-free neutral endeavour was abandoned within movements, though scientists continued to swear by it. This postpositivist phase, commencing three decades ago, has been marked by a quest for new origination, new questions, historiographies for studying the history of scientific and technological knowledge and, in no small measure, alternate sciences. The search for alternatives is of various hues, and one amongst these indubitably mirrors an Occidental neurosis concerning the reigning paradigm of the sciences—a neurosis that echoes in the Orient in the insertion of Islamic sciences, Hindu sciences, ethnosciences, etc. into the discourse of modern science.[19]

The romantic strain in India echoes a further disenchantment with science. One way of looking at the mobilization of the romantic as an ideology is to situate it in terms of the intractability of problems of the third world that has generated among third-world intellectuals a retreat 'into the womb of some primordial past' (Nanda 1991: 32). Within the world of science the malaise was to present itself in a symptomatology, which, to say the least, reflected the morbidity in the world of Indian science—characterized as enclaved and disconnected. However, what of the aetiology? The aetiology of the malaise, it may be conjectured, resided in four concomitant factors. First, the collapse of science's self-presentation as a neutral enterprise produced responses that wished to reground science in value or an egalitarian political order. Bernal, for one, thought that scientific workers' movements and organizations would successfully perform this task. Second, with the rise of the phoenix of American science in the post-War years, the marginality and dependency of India's scientific R&D system became increasingly visible. The Approach document of the National Council for Science and Technology attempted to redress the issue (*An Approach to the Science and Technology Plan* 1973). Further, scientometrics emerged as a pathological tool for monitoring the health of Indian science, the Citation Index serving as an indicator of health (Arunachalam 1979a, 1979b). The status of Indian science was cordoned off from its global context. The unwritten social contract between science and politics was substituted by a new arithmomorphism. Third, there was a clash of the Utopian image of science and the scientist (seen as sage and anchorite) with that of the scientist in an increasingly professional Indian environment.

In a recent neo-Gandhian reconstruction this anxiety is expressed as follows: 'Gandhi would have realized that what science needs least is money. What it requires is a more playful asceticism, a catalytic shift whereby science from being a career returns to being a vocation' (Vishvanathan 1992a). The critique of professionalization pleaded in actual terms for a restoration of the idyllic adventure science was possibly meant to be; which meant breaking out of a monolithic science in the hope of inaugurating an epistemologically and methodologically pluralist science.[20] Finally, there has been a retreat inward in the light of the disillusionment with the project of modernity, a disillusionment that is rendered diabolical by the conflation of the distinction between the project of modernity and the imperialism of the West.

The phenomena described indicates the broad contours of the counter-discourse. When discussing the romantic strain in India, there are two important strains that one must allude to. The first of these is situated in developments in the new anthropology of knowledge, carrying over in part the relativist programme into science studies, as well as a corpus of work that mirrors Western romantic articulations of science. Science was considered patriarchal, exploitative, reductionist, and an instrument for the propagation of Western hegemony (Nandy 1988). The writings of Feyerabend and Rorty have proved particularly influential on this count. However, Feyerabend (1987: 23), located as he is in the realist-scepticist perspective, had to issue a denial saying that he had been misread and argued that Western civilization was not pathologically violent and hegemonic. Amongst the romantics, the neo-Gandhians felt that the sources of an alternate science were to be found elsewhere, possibly in the traditions marginalized by the hegemonic march of 'Western science'.

The contributors to the *PPST* journal were seeking to resituate science within traditional Indian thought, drawing upon the cognitive aspects of the Indian systems of knowledge to reinvigorate the practice of science, making it more relevant and productive.[21] The notion of relevance could be construed to mean an attempt to situate the scientific tradition in a cultural context that is commensurate, in an organic (indigenist) way, with the socio-political trajectory. Surely, this effort has generated much knowledge of what appears to be lost to the self-consciousness of an imperialist Europe. The attempt to institute new research programmes has been dispersed and fragmented in epistemological terms. As the network of globalized science expands, other perspectives are naturally integrated in the received paradigm. However, the crisis of the nation state in the third world creates an ideological proliferation that either rejects or mobilizes science in ways that often endangers the universalist faith in science.

At the other end of the spectrum social movements that attempted to take science to the people date back almost to the first half of the 20th century. Popular writing on the sciences in Indian local languages celebrating Baconian science dates back to the first half of the 19th century (Habib and Raina 1989, 1992). However, the peoples' science movements broke out of the self-constructed enclaves of Indian science. Half a century ago the movement was restricted to the state of Kerala (Thomas-Issac and Ekbal 1987; Zachariah and Sooryamoorthy 1994); today they have mushroomed all over the country. Harking back to the early 1970s, a landmark in the changing perceptions of science, technology, and society, a whole generation of students trained in the sciences and exposed to the social sciences, as well as theoretical physicists and molecular biologists who had graduated from leading universities in the USA and located at some of India's premier science institutes, set out on a process of deliberation and action. The forum for this programme came to be the science movements (Giri 1992).

As a people's science movement, a number of positions had to be carefully accommodated. The point of departure for these movements was that to catalyse the democratic uptake of science it was imperative to enlist the masses to participate in programmes directed at their development. The process of social transformation was to be led by the march of science. Scientific activity was characterized by a *method* and a *temper*. This ideological programme found logical expression in a document referred to as the Scientific Temper Declaration.[22] The document was patently scientistic, reflecting the well-intentioned desire of scientists to hegemonize on epistemic grounds all forms of rationality and claims to truth. The only form of culture that the document acknowledged as exemplar was the scientific one. Any attempt to transform culture on such scientistic lines was fortunately shot down by a vociferous group of social scientists.

Since then the peoples' science movements have had to totter between three ideological formulations of the science–culture problematic: science in culture, science as culture, and science for culture. Only over the last decade has a formulation emerged among sections of the peoples' science movements that locate science as culture. Bernal continues to be an iconic figure within the movement, but his sway is more of a generational thing—for the immediacy of social action related to scientific activities would be governed by local conditions, experiences, and perceptions.[23]

If scientific controversies and the pressures from below have created the need for retrospection on the nature of science and technology, the social context within which the historian of science and/or technology is situated conditions the kinds of questions asked and the historiography that frames the historian's accounts. Developments in the history of science were informed by then pertinent questions, and in turn informed the movement itself. Studies on the internal history of science continue to appear in the *Indian Journal of History of Science*. Within the externalist studies of science I shall refer to three important strains: administrative, sociological, and socio-epistemological. According to *Bernal's weak thesis*, for science to serve effectively as a social institution it must be planned for in terms of human, financial, and material sources. This thesis provides theoretical legitimation 'for science policy doctrines in both East and West' (Elzinga 1988: 92–5).

The administratively-oriented studies are not concerned about the content of scientific knowledge or its epistemic manifestation. However, within the context of the history of science they have brought to light source material, and highlighted how particular scientific institutions were established. The emergence of these institutions under the umbrella of imperial rule and administrative constraints were discussed, and the debilitating effects of colonial policies for the advancement of science were studied. The modality for the transmission of scientific knowledge is essentially conceived of in terms of percolation models. The marginalization of local knowledges is explained as a consequence of the unfurling of imperial policies (Kumar 1995; Sangwan 1985).

The socio-epistemological strain is concerned with reinstating within STS discourse in India the idea of science itself as a cultural form. The switch from one knowledge system to another must then be explained in terms of the *dialogicity* between different systems of knowledge—involving producers, carriers, and recipients of knowledge. The notion of dialogue is predicated on that of shared meanings, in addition to the creation of new meanings and practices (Habib and Raina 1989; Raina and Habib 1990). Further, it opens up studies within third-world contexts to the recognition of the modalities of dialogue with the self (as in the indigenous systems of knowledge) and the other (modern scientific knowledge). The question that needs to be answered, lest the political edge of interpretation be lost, is when dialogue disguises confrontation. The distinction in emphasis between the two resides not in the divide between the epistemology of scientific knowledge nor the politics of knowledge.

In fact the socio-epistemological approach alludes to the interpenetration of the two.

Contemporary studies within the sociology of science have been concerned with a third-world pathology—the allegedly low productivity of science within the third world. In investigating these issues studies have ascribed part of the condition to the nature of the scientific community itself. Sociologists have attempted to decode the sense in which the scientific community exists in India. Does the community exist at all (Jain et al. 1992; Ramasubhan and Singh 1987)? Furthermore, within the postpositivist dispensation, as well as under the umbrella of postmodernist discourse, emphasis has shifted away from expert knowledge embodied in large scientific institutions to social movements contesting government legislation, wherein scientific or technological knowledge is implicated. Some of these studies are hopeful of a new origin in these endeavours (Giri 1992).

A CENTURY IN CONTEXT

A logical extension of STS over the past decade and a half is the application of the externalist razor to the historian (Graham 1985). When traversing the Indian terrain, we see how closely the concerns of Indian society, politics, and science are intertwined with questions posed by historians and sociologists of science. An acknowledgement of the fact is that when compared with the developed world India's percentage investment of its GNP in science and technology is marginal, yet over a century it has managed to create a huge scientific and technological establishment.

The inauguration, in institutional terms, of the history of science and technology goes back to the Orientalists; the discipline as recognizable today, bridging two distinct disciplines, history and science, came into being within a few years of the movement of Indian scientists to establish a research system in the sciences and technology. Late 19th- and early 20th-century science studies are marked by ambiguity about change, for colonial rule had resulted in the transformation of the continent. The need for change was perceived, but the issue at stake was what kind of change was to be instituted (Habib and Raina 1992). Discourses on scientific discourse were thus ideologically implicated in counter-colonial contests.

That India's first generation of historians of science were either scientists or science teachers is surely not coincidental. Within the network *of conservative modernizers* ways of manipulating the notion of progress—a notion that for the nationalists had to be defined differently from that of the colonizer—had to be worked out in order to legitimate their own agenda of progress and development. This undertaking aspired to render, as pointed out earlier, scientific activity as economically beneficial and morally worthwhile. Science was imbued for these interlocutors with an internationalism and neutrality—so much so that their scientific societies were the fora for articulating a politics of change.

However, in extant versions of the history of science, infused with the internationalism referred to earlier, they ran into Eurocentrism from Berthelot to Sarton. What was found problematic was the insertion within an internationalist context of a canonized European definition of rationality and universality. More importantly, the corpus of Orientalist literature on the sciences of India informed these readings in a non-Orientalist way, in that the interlocutors denied the East–West dichotomy. It is important to note that the corpus of works of some of the British and French Orientalists on the sciences was not always Orientalist in the Saidian sense of the term.

The decades shortly following independence, also labelled the Nehruvian era, could also be referred to as the high tide of modernization and development in India. These were the years of proliferation of India's S&T institutions: the atomic energy establishment was founded, along with the centres for advanced research in the basic and applied sciences, in addition to the laboratories set up by the Council of Scientific and Industrial Research. A number of heavy industries were established during this period, with a focus on the public sector. This was also the period when a number of large hydroelectric power stations came to be installed. Ensconced within this development programme was a view of science and technology that shared much in common with Bernalism.

In historiographic terms what was the order of the science-society relationship? First, as a newly independent third-world country, there was an attempt to reconstruct its 'self afresh', which in effect meant breaking with Orientalist and colonial reconstructions. A major shift in endeavour, departing from the religious positivism of the early 19th-century Bengali *bhadralok*, was the deconstruction of the notion of India as a spiritual civilization—of recovering non-idealist strains of thought from its past that would render future social transformism commensurate with a Brechtian vision of science.

The post-colonial world is replete with reconstructions. The reconstructions that we have been discussing had to polemically refute revanchist reconstructions of history and the history of science. In the process limitations in Marxist historiography were to surface. This was to produce both a conceptual and programmatic reconsideration in historical as well as history of science and technology studies. The period is also marked by the production of catalogues, bibliographies and source books, and translations of primary source material from the classical languages into English.

What has been referred to as the period of postpositivism and the crisis in development is characterized by both global and national crises. The broadly perceived notion among critiques of modernity is the construction of the West as the sole home and epitome of modernity. In epistemological terms this is announced in either the rejection of, or scepticism concerning, what an Indian interlocutor fondly refers to as 'The Immaculate Conception of Science', a conception that is premised upon the following two axioms: of the neutrality of scientific knowledge and the universality of this knowledge.

This produced a 'thousand flowers bloom' situation, marked by a proliferation of discourses concerning science, society, and development. We shall not summarize the historiographic disjunctions as has been done for the previous decades. On the contrary, we must ask what is it that forges them into a discourse. On re-examining the discourse of other social movements, the alternate science movements and the peoples' science movements, two essential features are evident. First, that a new discursive formation is gradually constituted—a formation wherein the discourse on science is transformed into a field of critical consciousness (Giri 1992). The formulation of cultural critique may also be seen as a modality of socially appropriating science and technology (Elzinga and Jamison 1986: 206). An additional feature shared amongst these critical voices is a symptomatic one—that of the dysfunction afflicting India's scientific and technological research institutions. Where they differ is in the diagnosis and prescription.

In any case, the academic responses resided in the institutionalization of policy studies in India. Programmatically, the voice of scientific authority was for once challenged. Since most scientific and technological R&D was government funded, the voice of the scientific establishment often resonated in the scientific community. The alternate science movements offered the contestatory force required to resist scientistic absolutism. On the other hand sections amongst the peoples' science movements, trapped between the welfare of the people and the steamroller of development, swearing by the ideology of science, were frequently compromised into becoming apologists of science. In this interregnum social studies of science sought to maintain a sense of balance between otherwise diametrically opposed programmes. As a result, whether these tensions came to be resolved or not, beyond distinctions of left and right, within movements and academic STS, science came to be relocated within the social and cultural bedrock.

NOTES

1. Barthes (1982: 155) has suggested that French Orientalism of the 17th and 18th centuries was a way of perceiving 'humanity at zero degree (Centigrade), which one nimbly grasps, to signify ... oneself'.
2. Kaneko (1987: 359) indicates the existence of homological phenomena in cases where 'an isomorphic state of mind or similar mentality can also be found in the semantic sphere of another culture'.
3. Dharampal (1971) has put together some British documents from the 18th century relating to the sciences of ancient India.
4. The accounts of the French traveller Francois Bernier (1989) served as an important introduction to the French Jesuits visiting India in the 17th and 18th centuries. Consequently, it could be asked as to how Bernier's (1992) *Abrege* influenced the disquisitions of the Jesuit fathers, for Bernier himself was a significant Gassendist.
5. William Ward, in 1821, in the defence of Serampore College was to write: 'And thus in this College, all that is good in Hindoo science, will be retained, native professors of the Eastern languages appointed, and European science engrafted upon the talents, the acquirements and energies of the natives' quoted in Kopf (1969: 261).
6. Romila Thapar (1993a: 30) points out that as opposed to the writing of British Indologists who were mainly administrators such as Lyell, Ibbelson, and Risley, 'French studies on society and religion in India came from professional scholars and sociologists.'
   For a review of French Indology, see Filliozat (1974), and a classic (Filliozat 1964).
7. One is reminded here of a perspicacious passage from Marx's *Eighteenth Braumaire* (1977: 10) which could just as well be read as a text on cultural appropriation: 'And just when they seemed engaged in revolutionizing themselves ... they anxiously conjure up the spirits of the past to their service and borrow from them names, battle cries, and costumes in order to present the scene of world history in this time honoured disguise and borrowed language'.
8. See Ray's (1932) autobiography to acquire an insight into some of his early research problems.
9. However, the partition of Bengal was an emotional schism in Bengali

selfhood. This, coupled with the emergence of a large intellectual proletariat, produced typically nationalist responses: militant political mobilization in the nationalist mould; and conservative modernization that sought to shift the focus of the debate to technical education and the onset of industrialization. For the nuanced nature of these responses, see Sarkar (1975b).

10. The next chapter discusses how Ray departed from the influence of Berthelot and Renan.

11. For two contemporary but divergent views, see Abrol (1995) and Visvanathan (1985).

12. Raja Rammohun Roy, in a sense the originator of the discourse on science and modernity in India, and amongst the founders of the Unitarian church in England, also the founder of a Hindu religious reform movement, the Brahmo Samaj, was simultaneously to take up cudgels with the Christian missionaries of Serampore, while writing a critique of the caste system (Roy 1821; Sarkar 1975a).

13. His other work includes Chattopadhyaya (1959, 1976, 1986).

14. For a review of India's economic and industrial base in the nineteenth century with a good bibliography see Chandra (1977).

15. Although this definition of the sense of history is located within the Marxist tradition, Kosambi is contesting Marx, for whom 'Indian society had no history at all, at least no known history. What we call its history is but the history of its successive intruders who founded their empires on the passive basis of that unresisting and unchanging society' (Marx 1853). The history of techniques was therefore a disciplinary field for the rejection of a certain construction of India, namely, that of a stagnant Indian social order.

16. Although for him the superstructure is not determined in any strong sense by the base, the nature of the determinations being far more polycentred, nevertheless a determination is postulated: 'The unlimited growth of superstition showed ... the necessity of the ruling class to subject itself to formal disabilities and restrictions in order to make religion effective in control of society. The advance of culture needs exchange of ideas, growing intercourse, both of which depend in the final analysis upon the intensity of the exchange of things: commodity production' (Kosambi 1985: 175).

17. For a discussion on problematic categories such as the Asiatic mode of production posed for historians at the time, see Chandra (1980).

18. For an exposure to this strain of research, see the issues of the *PPST Bulletin* published in Chennai since the 1980s. However, the *PPST*, while being critical of science and Western hegemony, would still work within the frame of modern scientific practice, which would mean working towards rendering their own work culture more efficacious. On the other hand, for a civilizational critique of modernity and

science, see Nandy (1980, 1988). For some of the older writing in the tradition see Alvares (1979). In addition to Nandy (1980), for another attempt to ground science on an alternate epistemology see Seshadri (1980).

19. A more recent phenomena has been the much-touted collection of rapid computation devices, put together by a school teacher, and being circulated around as Vedic mathematics. For a review see Dani (1993).

20. Visvanathan (1992b: 56) thus writes: 'the great folk science of Asia, central to food as to garbage, must recapture the imagination of any Third World science. The idea of bio-conversion also provides a pluralist touch.' See also Shiva (1988).

21. Some interesting work drawing inspiration from Panninian grammar and work related to natural language processing has appeared. For an interesting account of the evolution of Indian linguistics, see Singh (1986).

22. 'A Statement on Scientific Temper', *Mainstream*, 25 July 1981, p. 8. See Nandy (1981) for his 'Counter-statement on Humanistic Temper'.

23. For a source book on social movements and responses to development projects see Raina, Chowdhury, and Chowdhury (1997).

3

Prafulla Chandra Ray and Marcelin Berthelot: Chemist-Historians

This essay is divided into two parts, and seeks to explore the interplay between historical consciousness and politics in Ray's project. This reconstruction necessitates sensitivity to the history of chemistry as well as to political movements. For it is not coincidental that in a strange reversal of historical reflection, the historical narrativization of the antiquity of a discipline preceded the institutionalization of its modern variant in India. At stake is an interpretation of the transmission of scientific ideas viewed not merely as 'reception', but also of assimilation that leads to an interrogation for the social studies of science.

Traditionally, philosophies of science have endeavoured to undo the filigree of the context of justification from that of discovery, thereby de-emphasizing the shaping of scientific research programmes or the environment of research by the historical narrativization of science. This has largely to do with how 'influence studies' and historiographies premised upon the pre-Kuhnian philosophy of science, have conceived the project of history. From the vantage point of the politics of scientific knowledge, a shift in narratorial stance could be fruitful. The shift is a pressing consequence of the historiographic upheaval raging in post-colonial societies; and in the case of India it would possibly mean relocating the colonial enterprise as it was relevant to 'changing Indian concepts of cultural identity' (Lelyveld 1993: 665–82). This account is disposed to the consciousness and agency of the colonized, and the dialogue and sometimes militant opposition, which created the space for the ideological grounding of science in India.

Towards the third quarter of the 19th century, Indians at the periphery of the world of science were seeking out an exemplar for emulation in order to reduce the distance separating the centre from the periphery. While discussing the patterns of emulation at the periphery, Gizycki (1973: 474–94) is careful to point out that such programmes do not involve mere imitation but also adaptation of existing institutions, drawing them closer to models drawn from the centre.[1] To comprehend this process of adaptation in both institutional and cultural terms, we will propose a periodization in the stages of the adaptation and the role of individuals in this process.

The first stage is one where the autodidact has a significant role to play. The autodidact is situated in the indigenous systems of knowledge, and is pedagogically instructed in modernity. The autodidact is assigned the task of setting the terms of the dialogue through the activity of translation, which could be construed to mean actual translation—of textbooks of modern science into the vernaculars—which in turn requires the production of a cultural lexicon of metaphors and images that renders the world epistemologically refigurable in a frame that is recognizable and appealing. Further translation would also mean domesticating this knowledge through the definition of a grammar. The second stage is marked by a shift in location from the autodidact/man of letters to the professional, such as M.L. Sircar, founder of the Indian Association for the Cultivation of Sciences, and figures like J.C. Bose, P.C. Ray, and Hakim Ajmal Khan. These personalities hoped to salvage and revitalize those elements of the traditional systems of knowledge that are reconfigurable in the light of modern science. In addition to promoting science and instituting procedures for legitimating their ideological programme in scientific terms, the scientist became a partisan of the burgeoning nationalist struggle of the time. By 1914, through the efforts of the purveyors of science of the previous two stages, modern science came into its own in India. The first Indian Science Congress was organized, and the peers of Indian science graduated that critical number of students to ensure the further replication of the scientific research system. The choice of research problem in the exact sciences was divorced from its cultural grounding, and issues raging in the metropolises of science defined the research programmes—this is the third stage.

There is a remarkable difference in the three stages, particularly between the first two and the third. These differences may be seen as outcomes of the evolving nationalist movement, within which the scientific system is situated. According to Macleod (1987), British recognition of Indian independence did not come in 1947 but in 1914, the year the Science Congress was organized. The conflict between different systems of knowledge is the most conspicuous during the first two stages, which also makes this a rich area for the cultural studies of science. The tension between notions of modernity and tradition, both within colonial discourse as much as within the world of the colonized, is most acute in these stages. This tension is apparent in the ambiguity regarding both nationalism and internationalism. As indicated elsewhere, the first generation of Indian scientists, while positioned within so-called traditional societies, looked upon tradition critically, in order to revitalize these systems of knowledge and apprehended in science the social embodiment of the internationalist ideal (Habib and Raina 1989).

An issue of significance for the cultural studies of science is the examination of the 'process' of the acculturation/domestication of modern scientific knowledge in the indigenist idiom. Depending upon the perspective, some historians prefer the term 'cultural redefinition' (Metcalf 1986), a term encompassing as it does a civilizational finding relating to science, that denies percolation models sufficient explanatory potential and rejects the axiom that the method and verities of science are unattenuated by the cultural milieu it perfuses (Shapin 1983). By the late 19th century a section of the Indian intelligentsia had launched a project of reinventing modernity, and since science was the beacon of modern rationality, cultural appropriation necessitated the recovery of the idiom of rationality from within Indian society's cultural resources (Habib and Raina 1992).

The theme of this paper is the cultural redefinition of science in the second stage outlined earlier. A question germane to this stage of the cultural redefinition/appropriation of science is: when and under what conditions is the cultural redefinition of science possible? Further, could the choice of problems for scientific research or the structuring of a scientific research programme depend upon cultural and political considerations? To substantiate upon the nature of these connections two hypotheses will be examined:

Hypothesis 1 (H1): Essential to the cultural redefinition/appropriation of modern science is the establishment of a dialogue with the recipient culture's system of knowledge.

Hypothesis 2 (H2): The legitimation of science as a new knowledge form requires the possibility of deploying this knowledge for utilitarian ends as well as within emancipatory social movements.

This essay probes the corpus of work and activity of P.C. Ray, considered by many to be the founder of the tradition of modern chemistry in India.[2] In the first part of the paper Ray will be ideologically situated within his political context. A socio-epistemological reading of the dialogue Ray instated between the system of modern chemistry and the Indian system of alchemy follows. Ray, the chemist, launched the history of chemistry in India and laid the foundation of the social history of science. The second part of the paper suggests the interlocking of these two identities and is manifest in a chemical research programme marking the early years of his career. This informs our understanding of the cultural redefinition of science in non-European contexts, and the peculiar engagement of the ideology and the practice of science.

THE CHEMISTS' MILLENNIUM

A feature of science-based ideologies is their circularity, in that they aspire to become a science. But this becoming requires 'a constituted model of what science is' (Canguilhem 1988). The reconstruction that follows unfolds facets of 19th-century perceptions of science in India and the ideological legitimacy that was sought in science. For the student, Prafulla Chandra Ray, son of a traditional Sanskritist and Persian scholar, the world was already divided into two cultures. In fact, it was a world peopled by the icons of the European renaissance, and significantly by the 19th-century Bengali renaissance. The pantheon included figures such as Raja Rammohun Roy, the initiator of the project of critical modernity in India (Sarkar 1975a: 46–68), Debendranath Tagore, Keshab Chandra Sen, Akshay Kumar Dutt, the Bengal positivists, litterateurs such as Bankim, and radicals like Iswara Chandra Vidyasagar who had opposed equally strongly the teachings of Vedanta as well as those of Bishop Berkeley. These figures drew inspiration in part from the European renaissance and partly reclaimed the Indian past in very imaginative ways.

Committing Science to Nationalism

A typical conception of science in the late 19th century was that the wealth of nations was tied up with the state of development of the

institutions of science and with the capacity for technological innovation. In this conflation of science and technology, science came to
be coupled with nationalism (Dubos 1950; Paul 1985).[3] When P.C.
Ray returned from England, having obtained his doctoral degree
under Dr Crum Brown at Edinburgh, the nationalist movement in
India was still in its incipient stages. But modern science and technology was not the panacea for the shortcomings of Indian society.
More than nationalism, Ray's close relationship with the needs of the
peasantry 'and the masses in general' was reflected in his later years
in famine and flood relief work (Ray 1932: 40–1) and the mobilization
of the new knowledge in the task of 'development'.

His politics was at one level predisposed to the emancipation of
the 'oppressed' and, in the vocabulary of his times, opposed to
European rule over the Asian people. He writes that at the time of the
Russo–Turkish wars, he closely followed the heroic defence of Plevna
by Osman Pasha and Ahmed Mukhtar Pasha, for as an Asian his
sympathy was entirely enlisted on the side of the Turks (ibid.: 43).
However, his first act of sedition was committed as a graduate
student in 1885 at Edinburgh, when he participated in an essay
competition on 'India Before and After the Mutiny'. The essay competition was obviously organized by the prolocutors of British rule in
India. Naturally, he did not win the award, but the British sense of
fair play required that his essay be bracketed *proxime accesse runt*,
despite the 'bitter diatribes against British rule' (ibid.: 62). During
those years he appeared to go along with elements in the Indian
National Congress who promoted the doctrine of mendicancy to
make the British see reason. However, by 1905 the scene of struggle
had changed, and more militant forms of opposition had acquired
currency (Chandra 1969; Sarkar 1975a). Ray (ibid.: 63) observed: 'The
disillusionment was not long in coming. There is not in the history of
the world a single instance of a dominant race granting concession to
a subject people of its own free will and record.'

This transformation followed the partition of Bengal in 1905, but
sections of the Bengali scientific community subscribed to the programme of *swadeshi* (economic self-reliance), or constructive modernization, as different from the path of militant struggle adopted by
other sections of the community. This meant shifting the focus of
discussion to scientific and technical education appropriate to the
needs of Indian society and, in tune with the programme of industrialization, being drawn up by the burgeoning Indian industrial interests (Raina and Habib 1993; Sarkar 1975a). Ray was closely associated

with the National Council of Education (hereafter NCE), a pedagogic programme committed to *swadeshi*. In addition, scientific self-reliance, since the days of M.L. Sircar, required establishing a system of scientific research in India under Indian control and Indian management (Raina and Habib 1995; Sarkar 1946). Research was not part of the charter of Calcutta University till 1904, and Ray entirely sympathized with the programme of the NCE and was involved in it, so that students of science in India could devote themselves to 'original investigations', and that in the heated moments of 1907 was perceived within the community to be the need of the hour (Ray 1918: 23). Much later, in the 1930s, Ray was resentful of the fact that the British had failed to recognize the nationalist aspirations of the Indians, and invoked the republicanism of Voltaire and Rousseau, as did his 19th-century Indian forebears (Habib and Raina 1992), to caution the British of the inevitability of the realization of the aspirations of the Indian people (Ray 1932: 65).

Swadeshi, a programme aimed towards economic and scientific self-reliance, has been well documented in literature (Chatterjee 1986). One of the early tasks Ray set himself after he returned from England was the local manufacturing of some of the chemicals imported from England—some of which he borrowed from British pharmacopoeia (Ray 1932: 99–100).[4] The aim here is not to discuss Ray's programme of industrialization, but the relationship between his scientific research programme and his deliberations on the history of Indian alchemy during the 1885–1907 period. To fathom the latter it is imperative to examine his relationship with the nationalist movement. In the 1920s Ray was closely involved with the political fronts of the freedom struggle, and for all practical purposes was a practising Gandhian, much to the disappointment of his more radical students. In early 1901 Gopalkrishna Gokhale was in Calcutta and Mahatma Gandhi was his guest. Through Gokhale Ray met Gandhi and was actively involved in preparing the stage for Gandhi's first public appearance in Calcutta. The attraction Gandhi had for him was not merely political but extended to their shared asceticism (Ray 1932: 128).[5]

The Promise of the Millennium

It is interesting to note how Ray, on his return to India, chose his first research problems. These had to do with the application of chemical knowledge to the extraction of chemicals that were hitherto being imported from England. The findings of this research were published in the *Journal of the Asiatic Society of Bengal* and *Chemical Examination of Foodstuffs* between 1889 and 1894 (Ray 1932: 84). But the dream that inspired this research was that of Berthelot—a dream that prophesied that by the year 2000 AD all the necessary articles of food would be prepared chemically by chemists from the very elements and 'when foreign lands would not be worth fighting for, when wars and annexations would be things of the past as rich harvests would be gathered in the laboratory' (quoted in Ray 1918: 2).

The chief contribution of chemistry in attaining this millenarian vision derived from what Ray believed to be Wohler's inimitable invention of the field of synthetic organic chemistry (ibid.). And it was in this field that Ray hoped to create a degree of specialization in India, though he himself was trained as an inorganic chemist. Chemistry was for Ray, in his public incarnation, the only among the sciences 'calculated to develop the resources of our country and increase its wealth' (ibid.: 116–17). This raised the issue of whether chemistry qualified to be a science or merely technical knowledge concerning a particular domain. In promoting the cause of the Bengal Chemical and Pharmaceutical Works, and the need for a research facility that was imperative for this range of industry, Ray pointed out that even as a bread-and-butter science, chemistry had little need to be self-defensive, for as a science it did essentially gratify a fundamental human intention of unveiling the secrets of nature.

Prefigurations of the Historical Project

At a distinctly visceral level the empirical method of science found empathy within Ray's outlook since it further appealed to the Baconian metaphor of the scientist as a humble artisan probing the creation of divinity. The edge of this psychological imputation is blunted in that this Baconian vision echoes within the Bengali *bhadralok* discourse on science, but Ray was fairly acute in his deployment of the understanding that the cultivation of sciences must proceed alongside the application of the arts, if the transformative potential of science was to be realized (ibid.: 10). More importantly, this view was informed by the social history of science in Europe during the period 1780–1850. Naturally, he was drawn to the notion of the republic of science defined through the meritocracy of ideas and the radical political view this entailed. According to Ray (1932: 96):

If one studies the history of the progress of the technical arts and scientific inventions in Europe he will find solitary individuals working at a disadvantage and labouring under immense initial difficulties giving to the world the results of their indefatigable zeal and devotion, which have revolutionized the industrial world. They were almost invariably innocent of a high class education. A Le-Blanc dies in poverty in a foreign land ... James Watt ... of humble origin and yet struggling against the odds and surmounting insuperable obstacles.

The most remarkable feature of this genre of writing is the evangelical, missionary inflection that addresses the underprivileged amongst his readers through the Utopian promise of science as a worthy profession. The poverty of Bengal in the early years, and later the poverty of India, were to be his chief concerns; and he saw his research activity as a means of alleviating this poverty. The 'whole of Bengal is Nature's laboratory', a bounteous nature that scatters her gifts in profusion (Ray 1932: 88).

The early years of Ray's life as presented in his autobiography are fascinating, for the principal metaphor signifying the educational ideal is that of the humble artisan picking up the tools of the trade. The early years are years of apprenticeship, a sort of preparation for the project of the future, and in this his father (ibid.: 27) and the particular complex of Bengali culture played a significant role. This early exposure to the writings of Prafulla Chandra Bannerji, Ramdas Sen, and Rajendralal Mitra were imprinted in his mind; and Ray believed that his predisposition to antiquarian studies was acquired then, that this training was to stand him in good stead when he wrote his *History of Hindu Chemistry* (Ray 1932: 34).

In addition, the 19th century was also the century of comparative philology and the Orientalist project in India, and Ray drew inspiration from it. The Orientalists were an important influence in his early years, but later Ray radically departed from an essential Orientalist dichotomy. His philological mentor appears to have been his father, and the mature Ray, like most classicists, had acquired familiarity with some of the classical languages like Sanskrit and Latin, not to mention the staple fare of the 19th century, namely, French and German (ibid.: 37–8). This was essential to his historical project, and he sought the help of a Sanskrit *pandit* to decipher some of the treatises on Indian alchemy. We shall discuss the break with Orientalist scholarship that surfaced in the late 19th-century history of science writing that was announced through Ray's history.

In addition to the departure from Orientalism, there is in Ray's writing a recuperation of the compositeness of Indian culture that was being threatened under imperial rule. This compositeness of

Indian culture and the religious tolerance prevalent on the Indian subcontinent were effective rebuttals of Western constructions of India as the land of blind bigots. The rejection of this essentialization was the key to his disidentification both with imperialism and Orientalism. For he writes (1932: 66):

We find there is a tendency among a certain class of writers to single some of the worst type of Mohammaden despots and bigots, and institute a comparison between India under them and the India today .... It is forgotten that at the time when the Queen of England was flinging into flames and hurling into dungeons those of her own subjects who had the misfortune to differ from the dogmatic niceties, the great Mogul Akbar had proclaimed the principles of universal toleration, had invited the moulvie, the pandit, the rabbi, and the missionary to his court, and had held philosophical disquisitions with them .... Religious toleration, backed by a policy dictated no less by generosity than by prudence, was the rule and not the exception of the Mogul rulers.

This was an ideological battle with the British, the quotation being from an essay he wrote in England, but also signalled his distancing himself from a certain brand of Oriental scholarship.

Ray's turning towards the history of science is a turn towards novelty. The purpose of the history of science is more than pedagogic, for it directly impacts upon industrial culture; it is a way of analysing India's strengths and weaknesses, and is the bedrock from where future action must evolve. Consequently, he pleaded for the establishment of a school of synthetic organic chemistry in India. The supplantation of the natural dyestuff indigo by artificial indigo, following the synthesis of alizarin in the laboratory, had sealed the fate of indigo growers and the textile dyeing industry in India. India had to learn from the efforts of the Swedish chemist Berzelius, the French chemist Gay-Lussac, and the German chemists Wohler and Liebig. 'The history of the modem supremacy of Germans in the industrial world is the history of the triumphs achieved by generations of silent and patient workers in the laboratory' (Ray 1918: 4–5). This history of science was not a pragmatic inventory of the past that was to inform the present, but rather was on its own a text on the revolutions occurring in the realm of ideas, almost running parallel to the political revolutions that swept across Europe in the 18th and 19th centuries. Thus, the history of science became a legimatory trope for the articulation of the politics of change.[6] Ray compared the condition of chemistry in India in 1910 to that in England in the 1840s

(Ray 1918: 36). He read the history of science as a text on revolutionary change.

Furthermore, in articulating the need for chemical research in India, Ray emphasized that the radical programme of science ought to be mobilized within the framework of the resurgent political mood of the times. The politicized climate was receptive to revolutionary discourse; and Ray (1932: 37) portrayed his heroes as the martyrs of science: Bruno, Galileo and Paracelsus the 'ideal chemist ... an honest seeker after truth, who pursues knowledge for its own sake'. The history of the different branches of science provided ample illustration of the 'insuperable difficulties' faced by the votaries of science during the early stages (ibid.: 36). These difficulties do not merely reflect the obstacles in Ray's programme of founding a school of chemistry, but also the difficulties faced by the nationalists in furthering their ends.

An attempt will be made to situate Ray's historiographical deliberations in order to decipher how his reflections on history and science inform each other. For it is here, in humility, that this chapter makes a brave claim. A number of scholars (Bannerjea 1990; Chatterjee 1986; J.N. Ray 1961; P. Ray 1966; Roşu 1986; Sen 1986) have come close to deciphering this relationship, but have been prohibited on historiographic grounds from doing so. This taboo is rooted in a particular view of historiography and the history of science, and this reading suggests that Ray either suppressed his appreciation of this relationship or was a victim of the condition.

While the origins of the project can be traced back to the correspondence between P.C. Ray and Berthelot, an attempt will be made here to unravel the complexity involved in the undertaking. Ray adopts Berthelot as his exemplar, but Ray's history of alchemy departs from the Orientalist premises underlying Berthelot's project. If that be the case, Ray's history of Indian alchemy must be seen as a major historiographic landmark, and it could be reasonably argued that some of the concerns of the social studies of science in India in more ways than one prefigure in Ray's writing. Further, this project was foremost on Ray's mind during the two decades 1885–1905, at the end of which he again concentrated on research on synthetic organic chemistry, the founding of a school of research on the same lines, and the ensuing application of this knowledge to industry. Ray's institution of a dialogue between Indian alchemy and modern chemistry informs his ideas on industrial chemistry as well.

THE BEGINNINGS OF THE SOCIAL HISTORY
OF SCIENCE IN INDIA

I have special reason to look back to this period of my life with mingled joy
and delight. When you learn a new language, you have a new world
revealed to you as it were. (Ray 1932: 37–8).

It is likely that in the discussion that follows there will be a tension
between what is referred to as the 'member's account' and the
'stranger's account' (Shapin and Schaffer 1985: 4). What is offered
here is a 'charitable interpretation' (ibid.: 13). Thus, if it is less sympathetic to Berthelot, it is not to vindicate Ray, but rather to place
Ray's project at the centre of an alternative construction of the history
of science. During the years 1885 to 1915, Ray saw himself as essentially dedicated to chemistry: 'Chemistry claimed me exclusively as
her own' (Ray 1932: 67). He commenced his doctoral work in 1885
and retired in 1936. Upon completion of his doctoral work he con-
fesses to have become 'so passionately fond of chemistry' that he
decided to stay on in England for an additional year 'to pursue my
studies uninterruptedly to my heart's content' (ibid.: 68). This court-
ship with chemistry was to continue throughout his life, for he
perceived himself as an ardent devotee and student even at the end
of his career.

To his non-scientific readership the adventure of science and the
efficacy of its methods could be illustrated through personal example. From 1889, when he returned to India, until 1897 (Bannerjea
1990: 269) he did a great deal of work on the detection of adulteration
of edible fats and foodstuffs based on physico-chemical data, the
results of which he published in the *Journal of the Asiatic Society of
Bengal*. The autobiography mentions in graphic detail his pursuit of
what to the Western mind appears quaint, but in another sense is an
indication of the patient extension of the empirical methods of
chemistry to problems that were no longer central to the European
researcher. Furthermore, importantly, in these pages we see the insinuation of the idea that the wealth of nations flows out of the toil
of the scientist's laboratory.

The legitimatory agenda that underlies this persuasive writing
cannot be overlooked for he appeals to that realm within Indian
culture that could be cognitively commensurable with the programme
of modern science. Thus, he projects India as the tabula rasa for the
cultivation of the sciences for a millennium (Ray 1918: 37). However,
over the past 300 years the lamp of knowledge has glowed brightly

only in Europe, and it has been extinguished in India. The latter has come to pass because the schools of Indian atomism and logic, like Nyaya, have been eclipsed by the rising tide of idealistic philosophies such as Vedanta (ibid: 38). With the eclipse of the schools of logic and atomism, science on the subcontinent went into decline. If the decline were to be reversed, the light of science must shine again. In so reframing the argument, Ray was neutralizing the cultural import of science as Western, and thereby instituting the possibility of dialogue, rather than being pre-empted into a programme of hegemonic erasure of traditional knowledge by modern. In his research on the history of the medical sciences in ancient India, Ray observed the existence of elements of rationalism and the spirit of inquiry. In the canonical medical works of Charaka and Susruta, Ray noted elements of the empirical method and rational inquiry. Yet he did not refer to it as an empirical science, but one in which the practice of developed elements of surgery reveal the depth of knowledge gained from 'experiment and observation' (ibid.: 186–7).[7]

This very science, employing a biological metaphor, has like 'a potato taken kindly to the soil of Bengal' (ibid.: 44). The biological metaphor is suited to the cultural appropriation of modern science. For the last six centuries, Bengal had been the home of the Nawadwipa school of logic; and the efforts of the physicist J.C. Bose heralded a new age—the transition from the age of logic to the age of the physical sciences, in analogy to the European renaissance's break with the scholasticism of the medieval monasteries. But if science in India was to be revitalized, the question remained what sort of science was to be recovered, and what was its nature.

The history of techniques offered ample evidence of the existence of strong empirical traditions of metallurgy and other technical crafts in India. The Iron Pillar at Delhi was testimony to this tradition of metallurgy. But why did this empirical-technical tradition fall into decline? Ray addressed this question in the chapter 'Knowledge of Technical Crafts and Decline of the Scientific Spirit' in *A History of Hindu Chemistry* (Ray 1902, 1907). The divide between the arts, that was relegated to the lower castes, and the intellectual portion of the community precipitated a situation where 'the how and why of phenomenon were lost sight of, the spirit of inquiry died out ... (India's) soil was rendered morally unfit for the birth of a Boyle, a Descartes, or a Newton' (Ray 1918: 71–2).

Similarly, the decline of surgery followed the introduction of caste injunctions upon those who performed dissections upon the human body. The interdiction imposed on dissection presaged the end of surgery in India (ibid.: 191). The close link between practices and theoretical knowledge was a key concept in this historiography. Furthermore, other than enabling the secularization of the history of science in India, Chattopadhyaya (1986: 8) suggests Ray was a pioneer in the social history of science writing in the country. He was the first to recognize that the internal account is not sufficient to explain the dynamics of science and the unfolding of its history. In this historical project we witness an attempt: (*a*) to recover the past of science in India; (*b*) to legitimate science; but also (*c*) an attempt towards a social critique, for there was no use lamenting the past (Ray 1918: 123). The spirit of inquiry had to be reinstituted.

In 1894, while the idea for setting up the Bengal Chemical and Pharmaceutical Works was mooted, Ray had already begun to devote time to the study of Indian alchemy. By 1888 he was scanning Indian materia medica. Having carefully studied Udoychand Dutt's *Materia Medica of the Hindus* and Kannai Lal Dey's *Indigenous Drugs of India*, Ray collaborated with traditional scholars, the Kavirajas, to commence preparations of Kalmegh (*Andrographia paniculata*), Kurchi (*Holarrhena antidysentrica*), Vasaka syrup (*Adhatoda Vasica*), etc.[8] The programme was inspired in terms of the epistemology of modern medicine. For, as he writes: 'All that was needed was that their active principles should be extracted according to scientific up-to-date methods and that they should receive the imprimatur of the practitioners' (Ray 1932: 104). Ayurvedic medicine was to be reconstituted along modern lines. His fascination for pharmacopoeia spanned a decade and a half. The programme of revitalization did not merely extend to Indian pharmacopoeia, but also to agriculture, given the different soil types, crops, and agricultural practices in India. Ray (1918: 13–14, 66–7) also drew up a scheme of research priorities for disciplines where chemistry's impact was most significant.

The Bengal Chemical and Pharmaceutical Works was conceived as an industry with an in-house research laboratory that would develop efficient processes for manufacturing chemicals that were being imported from Europe. This required the skills of an analytical chemist, and Ray had trained several of them. Soon preparations like Syrup Ferri Iodidi, Spirit of Nitric Ether, and Tincture of Nux Vomica rolled out of his laboratory. About these first attempts he said, 'the very idea of locally manufacturing pharmaceutical preparations, which hitherto had to be imported, acted like a tonic' (Ray 1932: 106). The programme of economic self-reliance had to be premised upon a programme of scientific and technological self-reliance—in concrete terms this is what *swadeshi* was all about.

The history of chemistry was an economic morality tale illustrating the tremendous transformation of Western society. Wohler's synthesis of urea led to the forging of linkages between the research system and industry. The exemplar of this transformation was a German one. The case of alizarin was especially significant since it had hit the Indian indigo industry hard, jeopardizing the lives of thousands of planters and those associated with the textile industry. The establishment of a system of industrial chemistry was thus imperative for the health of the economy. Ray then drew up a plan for the development of the chemical industry, which commenced with the manufacture of acids and reagents absolutely essential for any chemical processing industry. Here, he proposed the installation of a sulphuric acid plant as the 'mother of all industries' (Ray 1918: 52–3). Based on the manufacture of reagents and acids, he planned the manufacturing process: starting from soaps to paper pulp to fertilizers and oils, and the whole range that follows suit.

Rewriting the History of Chemistry in India: French Inspiration

The industrial successes of chemistry, tied up with marked advances in chemical knowledge, generated in the latter half of the 19th century a significant body of scholarship on the history of chemistry. Through one such effort the history of chemistry in the modernist vein was launched in India. The renowned French chemist Marcelin Berthelot published his *Les Origines de l'Alchimie* in 1885. However, the work was incomplete in its treatment of Iranian, Indian, and Chinese sources. This hiatus was bridged by a three-volume opus, infused with the scientism of the *troisieme republique* (cited in Besson 1992: 141). The central thesis the work sought to place on irrefutable foundations concerned the origins of alchemy in ancient Greece and its diffusion in the Mediterranean basin, and subsequently to the Orient. The task necessitated a study of comparable practices and doctrines in Asia (Roşu 1990: 191).[9] Ray's examination of the development of Indian alchemy was at variance with Berthelot's account. The former's historical research in the area commenced sometime in 1894, for he mentions Berthelot's *L'Alchimie Grecs*, Kopp's *Geschichte*, and Udoychand Dutt's *Materia Medica of the Hindus* (Ray 1932: 115). The historical project was under way, for in 1896, Ray wrote to Berthelot, offering textual evidence that refuted Greek influence in

Indian alchemy. There is in Berthelot's *Les Origines* a passage that may have provoked a response from Ray. It reads: '*Le mercure ... joue un si grand role chez les alchimistes, est ignoré dans l'ancienne Egypte. Mais il fut connut des Grecs et des Romains. On distingue même le mercure natif et le mercure prepare par l'art, fabrique en vertu d'une distillation veritable, que Discoride decrit'* (Berthelot 1885: 231).

This Greek origin of the science of mercury, and the fact that *Les Origines* contains one reference to India regarding damascene steel and another imputing Alexandrian influence to an Indian alchemical text (ibid.: 140), may have initiated a project on the history of Indian alchemy to restore a sense of balance to the history of science (see Chapter 2). Ray's letter to Berthelot dated 1896 contested the claim that the Syrian Nestorians carried Greek alchemy to India and China (Berthelot cited in Ray 1932: 116).[10] This was the beginning of an exchange between the two, and the beginning of the history of alchemy in India undertaken by an Indian chemist. A few historic-graphical remarks on Berthelot would help highlight Berthelot's influence on Ray, and the difference that earmarks Ray's inauguration of the discipline.

Situating Berthelot

The history of this exchange, however, suffers from the drawback that all Berthelot's letters to Ray have been lost, but through the efforts of Arion Roşu (1986, 1990) some of Ray's letters to Berthelot have been reproduced. Roşu has identified the influence of the Orientalists on Berthelot's work, and the subsequent development of the French Indological tradition in the area of alchemy. While Roşu is primarily concerned with foregrounding Berthelot as the founder of a particular genre of the history of alchemy, the focus in this section is on Berthelot in relation to Ray. Berthelot and his friend, the Orientalist Ernest Renan, attended a course at the College de France offered by the philologist Eugene Burnouf (1801–52). And it is through Burnouf and Renan that Berthelot probably acquainted himself with the work of the Orientalists, and particularly that of Barthes, with whom Ray specifically disagreed.

Two essential features of Berthelot's history of alchemy were: (*a*) the developments in modern chemistry served as a bedrock for recovering the positive traces in alchemy ('*retrouver les traces positives*'), and (*b*) the histories of Kopp and Hoeffer served as models, but Lepsius' work on Egyptian metalwork and its adaptation in the hermetic school were of special interest (Roşu 1990: 189). In historiographic terms this meant that Berthelot was the first among the 19th-century savants to apply the exact sciences in order to gain an understanding of Egyptian metalwork (ibid.: 190). Roşu, with the historian's privilege of hindsight, pointed out the limitations in Berthelot's history. The first has naturally to do with the incipient stage of interdisciplinary research in the 19th century, reflected, according to Roşu, in the collaboration between a Hellenist or Orientalist ignorant of chemistry and a scientist lacking a sensitive insight into the meaning of obscure texts. Second, Berthelot inter-preted the mystical doctrines of the alchemists with excoriating contempt, which revealed his poor understanding of the same (ibid.: 191). The less edifying and more critical writing on Berthelot finds his treatment of alchemy equally problematic, for, according to Berthelot, alchemy was no more than a low-rate chemistry (Guillemain 1992: 110). Furthermore, while positivism inflected his writing, he adopted history as a shield to insinuate a quaint rejection of novelty in the name of modernity (Bensaude-Vincent cited in Besson 1992: 142). There are three interesting outcomes of his encounter with Ray: (*a*) it enlarged his vision of the origins of alchemy beyond the Medi-terranean; (*b*) while his history was philologically and historiographi-cally limited, it marked an epoch in the institutionalization of the history of alchemy; and (*c*) Palmyr Cordier, a doctor and Berthelot's contemporary, initiated the French tradition of the history of Indian medicine and the collection of alchemical manuscripts (Roşu 1990: 202).

The Berthelot–Ray Encounter: The Seeds of Disidentification

The year 1894, according to Ray (1932: 112–15), was an eventful year in his life, for not only did he move to his new laboratory, he also commenced his researches with redoubled effort, and began his researches into the history of chemistry. Of the 19th-century histories of chemistry, Kopp's *Geschichte der Chemie* had been his favourite reading since his days as a student pursuing a doctoral degree in chemistry at Edinburgh (ibid.: 110). In addition to the work of Kopp, the histories of Thomson and Hoeffer exercised an important influ-ence (Ray 1918: 75). He was introduced to metallic preparations in Indian traditions of alchemy through Udoychand Dutt's *Materia Medica*. While teaching at Presidency College in 1894, he chanced upon Berthelot's *L'Alchimie Grecs* (Ray 1932: 115).

Ray, the probable beginner in the field, wrote to the then pope of chemistry, saying that he had read the latter's work, but was aware of certain 13th-century Indian manuscripts that did not reveal Greek or Syrian Nestorian influence. Obviously, Ray had sent Berthelot his research papers recommending his scientific credentials. Berthelot responded: *'J'ai reçu vos recherches de chimie, qui sont fort interessantes et j'ai vu surtout avec plaisir comment la science avec son caractère universel et impersonnel est cultivée chez tous les peuples civilisés, en Asie, aussi bien qu'en Europe et en Amérique'* (ibid.: 116). Regarding Ray's countering his thesis, Berthelot asked: *'Mais je serais très curieux de connaître les traités indiens du XIlle siècle que vous me singalez. Ont-ils été imprimés?'* (ibid.: 116–17). Delighted by the nature of the response, Ray wrote an essay on *Rasendra Samgraha*, a 13th-century Sanskrit manuscript and mailed it to Berthelot. Berthelot reviewed it in *Journal des Savants* in April 1898 (Berthelot 1898: 227–36). The source of disagreement remained, as evident from the coolness of Berthelot's tone: *'D'apres ce savant, il existe des traités d'alchimie ecrits en Sanscrit remontant au XIII siècle et qui renferment des préceptes pour préparer les sulfures de mercure noir et rouge et le calomel employés comme medicaments'*. However, this riposte inspired Ray (1932: 117): 'I must write a history of Hindu chemistry modelled upon the exemplars before me.' The title of P.C. Ray's first piece on the history of Indian alchemy is interesting: *Materials for a Neglected Chapter in the History of Chemistry or Contributions on Indian Alchemy*, a manuscript numbering 43 pages. The title and its inspiration itself reflect its intent as a narrative of justice, of balance as an imperative. This negligence of a field so dear to Ray was not solely an outcome of ignorance, for though the Orientalists and other scholars had studied Indian philosophy, mathematics, astronomy, and to some extent chemistry, the latter had been neglected on account of its complex and technical nature (ibid.: 119).

The first volume of Ray's *History of Hindu Chemistry* appeared in 1902 and was highly recommended, given the fact that it was the first work of the kind. Most of the reviews acknowledged the original nature of the work, the facility with the classical languages, but more importantly, recognized the healthy distance maintained from 'stupid and senseless nationalism' (Hartog cited in Ray 1925: 107; Ray and Dutta 1911: 461). Gradually, the work became a standard reference in histories of pharmacopoeia and histories of chemistry: Svante Arrhenius, amongst many others, assigned priority to India in the use of mercurial and metallic drugs over Paracelsus, based on Ray's work. But more important was Berthelot's polite and almost gentle retreat in the review of the book in 1903 that appeared in *Journal des*

Savants: 'C'est un chapitre ajouté a l'histoire des sciences et des l'espirit humain, chapitre particulibrement util pour la connaissance des relations intellectuelles reciproques qui ont existe entre les civilizations orientates et occidentales.' The unidirectional model for the transmission of ideas had been transmuted into one of reciprocal exchange across civilizations. Prior to this, the realm of disagreement hovered around contesting interpretations of 'exchange': for Berthelot it meant an imparting from Greece, for Ray, a two-way process.

Nevertheless for Ray, Berthelot remained the guru, in characteristic Indian fashion: an exemplar of the chemist and the historian of chemistry. In 1907, when the second volume of Ray's *History of Hindu Chemistry* was published, Berthelot was no longer alive, and Ray dedicated it to the 'sacred memory' of the 'great savant and chemist' (Ray 1907). A quarter of a century later he referred to his communication with Berthelot as the turning point in his career as a student of the history of chemistry (Ray 1918: 75–6). But even so, while Ray did move the mountain, in the course of their correspondence, he conceded to Berthelot that India did not produce texts similar to that of Zosima and the Greco–Egyptian alchemists. In a letter dated 22 September 1898 Ray writes: 'I entirely agree with you that there is an absence of theories met with in the writings of the Greeks and the Arabians, and that we have to deal with only a collection of technical recipes and general principles' (reproduced in Roşu 1990: 203–4). It is only much later that Ray broke away from the scientism immanent in Berthelot's understanding of alchemy.

The Turn to the Social History of Science

In addition to revealing the significant contributions of Indian alchemy, Ray pioneered the social history of science in India, for he reckoned that the answer to the question concerning the causes for the decline of the sciences in ancient and medieval India may only be found within the politics of the community and the nature of knowledge claims they espouse. A detailed frame for the secularization of the history of the chemical sciences in India appears to have been developed by Ray.[11]

A brief discussion of the late 19th-century tradition of chemistry writing follows. Ray acknowledged acquaintance with Thomson's *History of Chemistry* (1830–1), Ferdinand Hoeffer's *Histoire de la Chimie* (1842), and Hermann Kopp's *Geschichte der Chimie* (1843). As Colin Russel (1988: 275) has pointed out, all three were practising chemists and, like all historians of chemistry at the time, Ray also confessed that Kopp was his starting point. Further, two points need to be mentioned in situating Ray among the late 19th- and early 20th-century historians of chemistry. 'Dotage theory' suggests that 19th-century historians of chemistry were 'retired chemists spending their declining years rewriting the history of their subject' in 'the best possible light', lacked the tools of historical analysis, and were Whig historians, in that they were concerned with what their generation considered 'successful science' (Russel 1988: 280–1). At the risk of producing a hagiography of Ray, it is essential to point out Ray's departure to show where he did not fit in with the elements of dotage theory. This is evident from the following:

1. His best work in chemistry was still before him. His project on the history of science began at about the same time as his work for which he achieved renown: in fact it may even be argued that the historical project was an indulgence of his youth.
2. He provided the first insights into the social history of science in India, and set the frame for subsequent externalist research. Along with Berthelot and Thomson (ibid.: 284), Ray shared a refreshing familiarity with original sources.
3. He overtly disassociated himself from Whig history (Ray 1932: 120), particularly of the British variety, and his own project sought to bring into the field of vision not what was already there, but that which had been lost.

The essential historical incommensurability underlying the Berthelot–Ray exchange is situated at the semantic (in the sense that each attributed a different meaning to the exchange of knowledge) and hermeneutic (how the alchemical texts were to be interpreted) levels. In addition, while Berthelot saw Greece as a singular source of influence, Ray's model was based on the polygenesis of knowledge. It is the hermeneutic argument that we will now turn to.

Beyond Berthelot and the Orientalist influence

Prior to Ray's work, most scholarship on the history of chemistry was confined to the work of Orientalists. Even so, the history of the cultivation of the experimental sciences was a neglected area. Orientalist writing was partially responsible for the notion that Hindus were a spiritual people whose writings were confined to transcendental teachings. This view Ray ascribed to the emphasis Orientalists gave to a study of the scriptures. Further, lest he might endanger his own interpretation by bending over backwards—the danger of conceptual presentism—Ray did not over-exaggerate the claim of the sciences of antiquity to be experimental sciences, but added that even in Europe the term was of recent origin (Ray 1918: 73).

Ray (ibid.: 74) axiomatically stated that experiments and observations 'constitute the fundamental bases of Sciences'. The alchemical knowledge of India is to be interpreted through this epistemic grid. This reading is reinforced by invoking passages from two important alchemical works, of the 13th or 14th century, *Rasendrachintamani* by Ramchandra and *Rasaprakasa' Sudhakara* by Yasodhara. Quoting from Ramchandra's work:

That which I have heard of learned men and have read in the Sastras–but have not been able to verify by experiment I have discarded. On the other hand those operations which I have, according to the directions of my sage teachers, been able to perform with my own hands—those alone I am committing to writing.

Or again: 'Those are to be regarded as real teachers who can verify by experiments what they teach—those are to be regarded as laudable disciples who can perform what they have learned—teachers and pupils, other than these are mere actors on the stage' (ibid.: 75).[12] In recovering the traditional sciences from the ayurvedic, tantric, and iatrochemical periods Ray lent credence to the aversion of science to scholastic arguments. Further, modern science could be legitimated without conceding to an ultra-radical epistemology or cultural xenophobia.

Ray clashed with his predecessors on two related fronts. The first is the academic milieu within which Ray's humanist peregrinations are initiated—the tradition that came down from William Jones. The second is through Berthelot. However, Ray soon felt the need to depart from the Orientalist reconstruction of the history of chemistry in India. He noted that 'very vague notions prevail among oriental scholars' on the subject. For the sake of brevity, we will rephrase Ray's main objections to Orientalist construction, and what he proposed instead:

1. Ray observed that the Orientalists painted the history of alchemy in India with a very broad brush. Barthes, for one, insisted that *Rasesvaradarsana* or the 'system of mercury' consisted of a strange amalgamation of Vedantism and alchemy (Ray 1918: 90–1). Ray's own history of alchemy in India identified three periods. The first is the ayurvedic, the second is the tantric, and the third is the

iatrochemical. In all three stages the place of mercury and its use in medicine is very distinct. These will be discussed later.

2. According to the Orientalists, the use of mercury and its compounds in alchemy in India was introduced by the Arabs. Berthelot was also of the view that from Greece, alchemy travelled with the Syrian Nestorians to the Arab world, and thence to India and China. Ray's reading indicated that there was Arab influence on Indian alchemy, but *rasayana* or the science of mercury was of Indian origin. The use of mercury in India predates the Arab influence. On the other hand, the works of Charaka and Susruta were translated into Arabic during the reigns of Khalif Mansur and Khalif Harun (ibid.: 92). There was disagreement between Ray and Berthelot on this count; and that remained unresolved, at least during Berthelot's lifetime. However, priority for the use of mercury and heavy metal-based compounds in therapy was attributed to India, even though Paracelsus was the founder of the practice in the West (ibid.: 92–3).

3. The limitations in the accounts of the Oriental scholars on Indian alchemy were a consequence of two historiographic presuppositions:

   (a) The Orientalists focused on Vedic sources that were more often than not repositories of ayurvedic medicine, that involved surgical practices and herbal medicine. However, to decipher the science of mercury (Sanskrit: *rasayana*), and a regime of therapy based on the prescription of metallic preparations, it was necessary to look at the non-Vedic sources that included Tantrism and Mahayana Buddhism (ibid.: 98).

   (b) Further, the Orientalists lacked a hermeneutic for interpreting the texts of alchemy. As Ray (ibid.: 91) put it: 'It is clear that the devotional formulae ... are here only a sort of jargon under which lies hid a radically impious doctrine.' This is a major departure of Ray from Berthelot, for in positing the need for a different hermeneutic he was liberating himself from the clutches of scientism and the more serious historical defect of presentism.

Ray's recognition of this problematic is evident in his appreciation of alchemical traditions outside India as well. In a popular article, certainly apocryphal by contemporary standards, he divulged, without any reference to his peers, the need for this break (Ray 1906: 237–8). In one breath, he denied a fundamental Orientalist dichotomy (Said 1978) of East and West, and recognized that in case the East is

East and the West is West, this self-similarity was not always so (Ray 1906: 237). Second, he distanced himself from those who perceived alchemy merely as vulgar charlatanism that seeks the conversion of base metal to gold. On the contrary, the Tantrists in India, the Rosicrucians in Europe, and Paracelsus, the 'sage and seer of Hoenheim', were all seekers of truth (ibid.: 238). Patanjali and Nagarjuna in India and Paracelsus were 'dreamers, mystics and naturalists combined in one' (ibid.: 238). Clearly, he had departed from the rationalist abuse of his forebears, and dissociated himself from the scientism of the third republic in favour of a more contextualist interpretation. As a result Ray had proposed a historiographic frame for interpreting the history of alchemy that brought to bear a new understanding of the alchemy of India. The crux of Ray's historiographic disidentification with Orientalism was that the exclusion of elements of a heritage was an essential ingredient of the politics of knowledge. While this dimension was never explicitly stated, Ray's participation in the nationalist struggle, and his unhappiness with the extant historiograhies of alchemy are but manifestations of the politics of knowledge.

THE CONJUNCTURE OF SCIENCE AND HISTORY

Dans la proportion même ou l'historien des sciences sera instruit dans la modernité de la science, il dégagera des nuances de plus en plus nombreuses, de plus en plus fines, dans l'historicité de la science. La conscience de modernité et la conscience d'historicité sont ici rigoureusement proportionelle. (Bachelard 1971: 201)

Scholarship on Ray indicates that between 1895 and 1910 he was simultaneously committed to three distinct streams of activity (Chatterjee 1986; J.N. Ray 1961; P. Ray 1966; P.C. Ray 1932; Sen 1986). Chatterjee (1986: 13) typifies the quizzical response of all three: 'How he managed to combine the three different streams—the continuous devotion to industrial chemistry, the researches in pure chemistry and the deep studies involved in deciphering the old manuscripts for the book on Hindu chemistry—all demanding full time attention will puzzle anybody.' An attempt will be made to establish that these apparently independent projects mutually informed each other. To grasp the manner of this informing it would be essential to reopen the black box of 'problem choice' in the 'pure sciences'; for it is likely that within the domain of the cultural studies of science a shift from the epistemic domain would throw up different interrelated regimes.

Sen (1986: 61), a student of Ray, expresses the uneasiness concerning

Ray during these years: 'Why did P.C. Ray confine himself to the investigations on nitrites from 1896 to 1912 particularly when major discoveries were being made in other areas of chemistry and physics during the late nineteenth and twentieth centuries'. This he considered an indulgence, without explicitly saying so, for he proceeds to explain: 'The reason might have been that Ray was involved in other activities besides research .... From 1896 to 1906 or even later he was busy collecting and studying materials for writing his magnum opus, the History of Hindu Chemistry' (ibid.: 61). He then points out that this was probably of 'equal importance' as his work in the laboratory. But despite the qualification, however obliquely judgmental, this is not the point at all. Recognition as a chemist came P.C. Ray's way in 1896, for research undertaken during this decade. The point that Ray's chemical problematic appears quaint from a later day perspective is a consequence of historical (chemical) presentism, and partly due to the inability of members of the community of chemists to visualize the possible relationship between a scientific research programme and a historical project. This image of the independence of science from a reflection on the past of the discipline was pervasive in the early decades of the 20th century, though not necessarily in the mid-19th century, wherein the history of science was considered part of science.

A Summary of Ray's Chemical Researches

In 1877 the University of Edinburgh awarded P.C. Ray a doctorate for a thesis on 'Conjugated Sulphates of the Copper–Magnesium Group: A Study of Isomorphous Mixtures and Molecular Combinations'. Though an inorganic chemist by training, his research problems in the subsequent years necessitated forays into physical and organic chemistry. The second half of the 19th century was a period when physical chemists in Germany sought to professionalize and establish their discipline as an autonomous sub-domain alongside pharmaceutical, medical, and analytical chemistry (Hiebert 1982: 97). During this period a number of research areas were brought under the jurisdiction of physical chemistry, but the central problematic concerned the study of chemical change, which effectively meant obtaining the position of equilibrium, and calculating the speed of chemical processes (ibid.: 101).

From 1894 to 1896 Ray undertook analytical investigations of Indian rocks and ores to fill the gap in Mendeleev's periodic table. While pursuing this task, he, as the official account evokes serendipity, synthesized mercurous nitrite in 1895. This is not the place to discuss the sense of 'discovery', but what is worth noting is that a two-faced 'science of mercury', in the alchemical and modern chemical incarnations, occupied Ray during the next decade. During these years he pursued the study of mercury compounds, followed by studies on nitrites and hyponitrites of other metals, including alkaline earths in the pure states. This decade may be considered the period when Ray, in terms of his chemical repertoire of skills and techniques, switched from purely preparative inorganic chemistry to the measurement of physical parameters and the determination of the physico-chemical properties of compounds (Sen 1986: 41). By 1907 Ray had more or less initiated a tradition of researches into inorganic and physical chemistry, and had moved away from his obsession with what is referred to as his 'science of mercury' years, to the founding of the school of chemistry and the research programme in organic chemistry. We refer to the 1895–1907 period as the 'science of mercury' years because they weave the effort of a decade into a coherent thematic whole, embracing in the process two epistemological projects.

Problematizing the Science of Mercury

We will now discuss the close relationship between the actual research done by P.C. Ray on mercury and heavy metals, and the edification of this work through his research on the history of alchemy in India during the medieval and ancient periods. This will provide us an instantiation of the conditions under which, Bachelard (1971: 202–3) insists, the history of the sciences can have a positive impact on scientific thought. But here we extend the Bachelardian framework a little further and suggest that these two activities mutually informed each other. This mutual informing is realized through the conjuncture of three distinct orientations. First, Ray's own predisposition to pharmacological chemistry and a specific programme initiated by his thesis supervisor at Edinburgh, Crum Brown, who along with Thomas Fraser founded the branch of pharmacology dealing with the constitution of drugs and their therapeutical properties (Ray 1932: 60). Second, his investigations revealed that the place of mercury and mercury-based compounds in Indian alchemy was unique, and he intended to decipher the nature of its use, preparation, and efficacy. Third, in 1895 Ray commenced his research afresh on the problem of assigning a place to mercury and some of the heavy metals in the periodic table.

The idea is not to suggest a causal relationship of the type where a set of factors precipitates a particular set of actions or a kind of activity. For even Ray and his students would shy away from such a suggestion—there does not appear to be an insinuation of this nature in any of their published writings. Nevertheless, this reticence was an essential element of the regnant historiographies that see the history of science as offering an imaginative account of already accomplished science, of buttressing the scientific communities' preoccupation with history. The Bachelardian perspective may provide another way of viewing the relationship between the history of science and contemporary scientific practice. We hope to instantiate in the subsequent subsections the process of the two mutually informing each other, revealing the modality of dialogue between the ancient sciences and the modern—a dialogue whose nature was transformed by the time Ray had retired from professional activity.

There is another way of looking at the same phenomena. In a recent work on J.C. Bose, Subrata Dasgupta (1999) evokes some of the recent research in the area of cognitive science, in particular that of Howard Gruber, and suggests that 'the work of a creative individual typically entails a *network of enterprise*'. This network refers to the group of related projects taken up by a scientist at a given time (ibid.: 119, emphasis mine). If the network of enterprises constituting Bose's project 'had a common goal: to demonstrate the essential connectedness and unity of diverse phenomena in the natural world' (ibid.), Ray's network was to investigate the place of mercury in diverse chemical traditions, and thereby to reveal the therapeutic properties of mercuric compounds, if such they were. While Ray's work was less metaphysically oriented than Bose's, the former's three projects taken together were psychologically significant as in Bose's case in providing a creative scaffolding for their endeavours (Dasgupta, 1999: 120).

The Science of Mercury: The Historian's Account

The first part of this paper described the events that led Ray to the study of Indian alchemy, the commencement of his researches in the area, the presentation of India's unique contribution to alchemy, and the disagreement between him and Berthelot. This agonistic contest, in retrospect, appears to have been settled in Ray's favour. More specifically, Ray was committed to the polygenesis of alchemy, and priority in the use of metallic preparations in therapy: the latter did not amount to asserting that this influenced Paracelsus a couple of centuries later. For Berthelot the project on the history of alchemy was one in self-congratulation, through the imputation of a cultural unity that was traceable back to the Greeks. The disagreement became most obvious in the second part of the manuscript Ray had mailed to Berthelot (Roşu 1986: 72). Again, in a letter dated 23 June 1898, Ray informed Berthelot that he had consulted some more texts on Indian alchemy and had found in them a process for removing the liquidity of mercury by titrating globules of mercury with a vegetable extract and heating it in a closed retort (Roşu 1986: 74).

The utility of alchemy lay in its connection with medicine. In the history of Western medicine, Paracelsus was credited with the introduction of metallic derivatives in therapy, and the use of mercury for treating syphilis (Knight 1992: 21). This is considered a major revolution in the history of medieval medicine. However, as Knight's history of ideas in chemistry informs us, one of the fundamental problems facing early alchemy was to make base metals appear golden, and mercury and sulphur were seen as the key. Mercury and sulphur, and salt, the third conservative element of the triad, were observed to be the constituents of all metals (Knight 1992: 16). In his first paper on *Rasendra Samgraha*, Ray had reported that the Indian alchemists knew how to prepare black and red compounds of mercury and sulphur that were used in medication. While Berthelot believed that this naturally confirmed Greek influence, Ray insisted on independent discovery.

There is no reference to the science of mercury in ancient Indian medical literature such as ayurveda (Ray 1918: 77). Ray not only initiated the modern history of alchemy in India, but also in this history his specific contribution, other than the historiographic one, lies in the disclosure of the evolution of the science of mercury. A separate discussion on this history is warranted. Ray established that there are two important phases in the use of mercury and metallic compounds in Indian alchemy. In the first phase, which corresponds to the tantric period, the discipline was documented in canonical works like *Rasaratnakara* and *Rasarnava* (literally meaning the 'sea of mercury'), mercury-based compounds were sought out and prepared to serve as the elixir vitae.

However, during the period which Ray called the iatrochemical period, whose alchemical knowledge is contained in such works as *Rasendrachintamani*, mercury and metallic preparations were used as accessories in medicine, as opposed to surgery and herbal therapies (Ray 1902: lvi). By the 13th and 14th centuries this knowledge ('the employment of mercury and metals') was exclusively introduced in medicine: and hence Ray ascribed to the term *rasayana* the connotation of the science of mercury (Ray 1918: 78). A century later inorganic/metallic compounds were elements of a medical practice that 'reacted upon the age in giving fresh impetus to the study of chemistry' (ibid.: 86).

The disagreement with Berthelot remained since the unani tradition as evident in India showed a strong aversion to the utilization of metallic drugs in medical practice (ibid., 92–3).[13] Further, as pointed out earlier and reiterated by Ray, mercuric/mercurous and other metallic preparations were first pressed into European pharmacopoeia by Paracelsus in the 17th century. Thus, India retained for Ray priority in the use of mercury-based drugs in medical practice as the tradition was possibly unique on this count. Though Ray's history appeared in 1902, the textual evidence appeared to have settled the issue for him. Nevertheless, in terms of historical and political consciousness, there remained a challenging epistemic obstacle to overcome, namely, Berthelot's presentism. While Ray was conscious of it, abandoning it willy-nilly would have meant writing away his own legitimatory agenda.

Mercury in the Periodic Classification of Elements: The Chemist's Account

If Berthelot was an important source of inspiration for Ray, the other was the great Mendeleev. In one sense Ray's narratorial range is influenced as much by Mendeleev's *Principles of Chemistry*, a work that exemplified the 19th-century chemist's landscape; for Ray considered this work a 'classic in the domain of chemical literature' (Ray 1906: 461). Bensaude-Vincent (1986: 3) has highlighted two features of Mendeleev's *Principles of Chemistry*: the projection of chemistry as a science firmly established on 'principles derived from experiment'; and the facile mobility in discussing problems of physics and chemistry to the problems of Russia's economic development, the inability to separate the 'future of chemistry from the future of Russia'. Ray reckoned with the idea that the periodic system was a break with chemistry's past that rendered chemistry a 'rational and predictive science'. The 19th-century chemist, in the wake of the new formalism, was propelled into the role of an adventurer seeking out new elements that could fill in the gaps in nature's ordered schema. Ray's (1932: 1) autobiography begins with the exclamation: 'I was born on August 2, 1861. This year ... is memorable in the annals of chemistry for the discovery of thallium by Crookes.' Wherever he discusses his vision for chemical research in India, he indicates that it would be essential to scout the Indian topography for possibly new elements, compounds, and ores. By 1894 he had a remarkable collection of mineral specimens obtained through his friend Thomas Holland from the Geological Survey of India (Chatterjee 1986: 13).

After obtaining his doctoral degree, Ray spent sometime overcoming his self-professed inadequacies in 'organic chemistry', as the benefits accruing from the pursuit of research in organic chemistry would herald the arrival of the millennium (Ray 1932: 71). This was also the time he began contemplating his return to India. He returned to Calcutta in August 1888. He mentions that his research in chemistry began afresh in July 1894. His research started off with the attempt to analyse rare Indian minerals in the hope of discovering *two new elements* that would fill the gap in Mendeleev's periodic table (ibid.: 113). Evoking chance, Ray says that in the process they found that mercurous nitrite had been synthesized. This was followed by the synthesis of a large number of mercurous compounds. The first paper was published in the *Journal of the Asiatic Society of Bengal* (Ray 1896a: 1–9). The article was also mentioned in *Nature* (1896), where it was pointed out that a paper of this order had no business to appear in the *Journal of the Asiatic Society of Bengal*. His subsequent work (Ray 1896b: 365; 1896c) was published in the prestigious *Proceedings of the Chemical Society* and the *Journal of the Chemical Society*, which ranked high in the profession. The encomiums came in from his teachers and other renowned chemists of the time such as Roscoe, Divers, Berthelot, Victor Meyer, and Volhard (Ray 1932: 114). In the same breadth, Ray mentions something else that was emerging at about this time. He talks of his reading Berthelot's book and his correspondence with him on *rasayana* (the science of mercury—a concordance of two mercury-related projects, but thematically different spheres of inquiry). Nevertheless, while he chanced upon the synthesis of mercurous nitrite, his investigations on the history of Indian alchemy had revealed to him the centrality of mercury in the medieval period. However, it appears that his scientific credentials concerning mercury were legitimating his historical claims about the science of mercury in India.

Ray's papers on mercury and the heavy metals focused on assigning a place to mercury in the periodic table. The problem was that mercuric compounds, or compounds of dyadic mercury, bore properties that were closely analogous to those of magnesium, zinc, and cadmium, and thus warranted a place in Group II of the periodic table. On the other hand mercurous compounds bore a closer analogy to silver; mercurous nitrite behaved analogously to silver nitrite and was as stable. The outcome of his researches on heavy metals and the detailed studies on mercury led to the inclusion of monad mercury at the bottom of Group I of the periodic table, while gold moved to Group VIII. Thus, univalent and bivalent mercury are very different elements, the former being closely related to elements like silver. This duality of properties of mercury makes it 'comparable with thallium' (Ray 1914: 85). Two decades later, Ray summarized his researches of those years in *The Chemical News*, then edited by Crookes, where Ray suggested that the fall out of his work was to assign a place to mercury in the periodic table. Nowhere in this account is there a reference to any problem outside the scope of what fell within the purview of discussion of inorganic chemistry.[14] The anguish and joys of those fateful years, the intersection of a plurality of narratives, are totally absent in the account. The professionalized presentation of the scientist is complete. By 1913 nearly 40 papers had been published in related areas and a research group had been constituted. Other detailed works are available on Ray as the founder of the tradition of synthetic organic chemistry in India (Guay 1986). But then, that is the story of the later Ray.

The issues raised above relate to a point made by Bachelard and later by Canguilhem: that if the role of an epistemological approach to the history of science is to shift the focus of interest from the 'history of science to science as seen in the light of history', then Canguilhem (1988: 3) asks, 'does this science of the past constitute a past for the science of today?' It is this inversion of the role of the history of science, whether consciously or otherwise, that marks the conjuncture of the allegedly disparate projects of P.C. Ray during the decade and a half that weaves them together. In so interrogating the past in the light of the present, the past of science was till the end of the 19th century, shall we say, present continuous.

In the work of Ray this conjuncture is constituted through the elaboration of two sub-themes: as a chemist assigning mercury a place in the periodic table; and the relationship between chemical properties and physiological properties. His enterprise as a historian of chemistry seeks to elucidate the place of mercury and metallic compounds in therapeutic practice. Here we see the intertwining of two strands in the early Ray's life: on the one hand he is the modern chemist studying mercury, on the other, in studying *rasayana* (the alchemical science of mercury) as a *rasasiddha* (Sanskrit equivalent of an alchemist), Ray assumes the identity of a *rasasiddhapradayaka*, a term that Ray translated for his non-Indian audiences as an expert on the science of mercury (Ray 1918: 78). In this twilight zone, astride two distinct epistemological programmes, the travails of the early Ray cease to be quixotic and acquire a renewed coherence.

THE END OF INNOCENCE AND THE COMMENCEMENT OF 'THE SCHOOL OF CHEMISTRY'

By 1907 Ray's 'science of mercury' years were beginning to come to an end (see Figure 3.1). He reckoned with the fact that in the interim period a whole new world was being instituted by chemists and atomic physicists: the Curies had completed their studies on radium; Rayleigh and Ramsay had discovered the rare gases argon, neon, xenon, and krypton; and radioactive properties were being studied by Rutherford and Soddy. By 1905 Ray felt that too much had happened; his historical works were now on the back burner: 'I was buried in my researches on the chemical knowledge of the Hindus of old and therefore losing touch with the modern world' (Ray 1932: 122). A phase now begged closure, and efforts were under way towards instituting a full-fledged tradition in synthetic organic chemistry.

In 1930s, on the eve of his retirement, his students remarked that: (*a*) Ray's theoretical researches in chemistry led him on to the application of this knowledge to harnessing the material resources of India, and (*b*) his *History of Hindu Chemistry* helped build bridges with the past, so that modern Indian researchers could turn back to Charaka and Nagarjuna with pride (ibid.: 189). But by now, ensconced within the culture divide of the world of science, the underlying unity of Ray's project was lost to its times and the age thereafter. However, the success of the project lay in that a tradition could turn upon itself without much diffidence. The space for modern chemistry had been gained without the need for external intervention or other intrusive strategies. What we have tried to show is that during these years his research in pure chemistry and his investigations on the history of Indian alchemy very closely informed each other. Furthermore, this process of informing lay at the intersection of the deliberations on nationalism and history on the one hand, and a dual project related to the relationship between chemical constitution and physiological response on the other.

As for Ray himself, the historical project was also political—in as

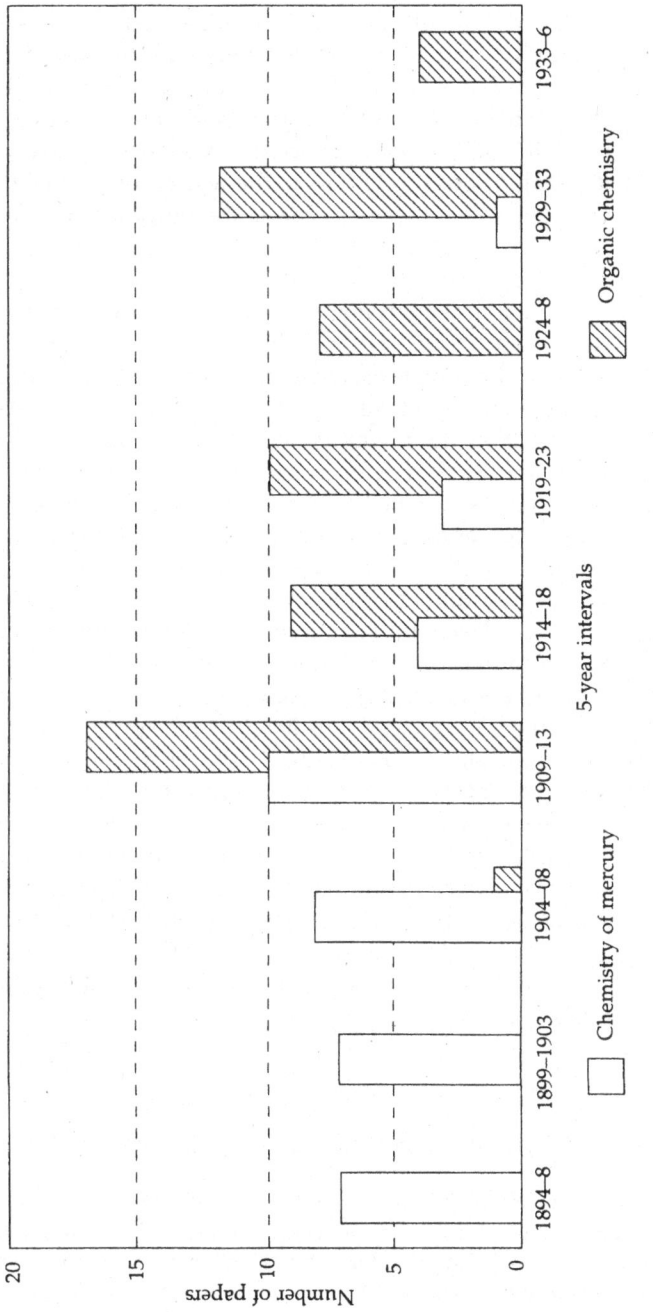

Figure 3.1: Ray's Publication Profile

much it countered Orientalist and colonial definitions—and legitimatory: 'Hindu Chemistry ... waited long and patiently for an interpreter. I thought I owed a debt to that great nation to which I am proud to belong.' Further, 'I implore you to take to its pursuit and I hope that you will justify by your work that you are no unworthy successors of your glorious forefathers in the world of learning' (Ray 1918: 102). On the surface these appear as devoutly nationalist remarks, but when read together the legitimatory agenda becomes obvious. In fact, the psychologist Girindrasekhar Bose, a contemporary of Ray, had the following remarks to make when promoting his discipline: 'India's ancient learned men had a genius for introspective meditation and the Indian psychologist has that heritage. In this respect, he enjoys an advantage over his colleagues in the West' (quoted in Nandy 1995: 143). Nandy sees Bose's remark, similar I would say in spirit to Ray's, not as an instance of 'facile expressions of nationalism', but as 'a construction of the past oriented to a preferred future, and serving as a critique of an imperfect present' (Nandy 1995: 143).

There have been a number of responses to Eurocentrism within the domain of STS studies in India. Studies on the cultural appropriation of science in modern India, a field still in its infancy, cannot escape returning to the years 1870–1920, when the conflict between different knowledge systems was at its height. In the present context, Ray's was an important conflict-ridden cultural dialogue, across distinct knowledge systems. Sociological approaches to the history of science have recognized that the history of science, more than any other genre of history writing, has been the most vulnerable to the Whig conception of history (Shapin and Schaffer 1985). The present study has sought not to revive Ray,[15] but to understand a specific cultural response to modern science (Ray's work being one in a larger cultural formation) that was to pave the subsequent trajectory of science in India.

NOTES

1. However, Gizycki is concerned about the centre and peripheries in Europe.
2. P.C. Ray, in an article written with Bidhubhusan Dutta informs us that the 'first chemist of Indian blood was Laurenco, a fellow student of Henry Roscoe and a pupil of Bunsen at Heidelberg in the early fifties of the last century', who for lack of opportunity in India went on

to become a professor of chemistry at Lisbon (Ray and Dutta 1911: 460–2).

3. Adas (1990) has discussed the emergence of technology as a measure of a nation's development in the age of late colonialism.

4. Mobilizing contributions locally, he set up the Bengal Chemicals and Pharmaceuticals, with a research facility to pursue the development of processes for the local manufacture of pharmaceutical products. The establishment of the Bengal Pharmaceuticals is a separate issue that will not be discussed here; suffice it to say that pharmaceuticals became his new-found passion (Ray 1932:103).

5. On P.C. Ray's 70th birthday, Francis Conan, Professor at the University College of Science, was to refer to Ray as one who 'never asked much for himself, living always a life of Spartan simplicity and frugality', calling him the 'Saint Francis of Indian Science' (quoted in Ray 1966: 72).

6. Lavoisier in France opened up a whole new tradition of chemistry that gave rise to the tradition of Gay-Lussac, Dulong, Thenard, Ampere, Arago, and Chevreui. In Germany Wohler and Liebig played the same role by inaugurating the discipline of organic chemistry. England drew its inspiration from Priestley, Cavendish, Dalton, and Davy, but was slow in following suit in the wake of Germany and France (Ray 1918: 34–5).

7. Subsequent histories of the sciences in ancient India have taken off from this observation of Ray. Debiprasad Chattopadhyaya (1979) refigures Ray's thesis in a Marxist framework, to point out that of the sciences of the ancient period in India it was medicine that best qualified as a science in the modern sense of the term. Further, Ray pointed out that the tradition fell into decline when the sciences were divorced from the crafts: the divide between theory and the technical crafts.

8. In 1910, by which time he had already published two volumes on the history of Hindu alchemy, with K.C. Kabiratna he brought out a joint translation of the Sanskrit text, *Rasarnavam* (Chatterjee 1986: 13).

9. Roşu further points out that in retrospect Berthelot has been proved wrong in imputing Greek borrowing in Indian and Chinese alchemy.

10. Ray's letter dated 9 February 1898, the longest letter from Ray to Berthelot, states that his historical researches have been interrupted by his scientific work on mercurous nitrite, etc., and that among other things, he desires informing the European scientists of the indigenous origins of Indian alchemy (xerox of the letter reproduced in Roşu 1990).

11. Marxist writings of the 1950s and 1960s are elaborations of these insights; while Ray himself was not within the Marxist framework, he was familiar with the work of the first Marxist history of chemistry authored by Carl Schorlemmer. This familiarity is not coincidental, for as Tucker (1993: 643) has pointed out in a more general context, much

of the 'initial theoretical incentive for the development of economic and social historiography was Marxist'.

12. This quotation appears on the cover of the 1897 manuscript.

13. This debate was of a foundational nature, in that the exchange provided the occasion for the founding of French Indological studies on alchemy and ayurveda, but has been totally ignored by even more recent works, for example, Charles Leslie's (1976: 357) *Asian Medical Systems*.

14. Between 1895 and 1907 Ray worked on the following. He isolated crystals of mercurous nitrite by the action of dilute nitric acid containing 13.14 per cent N_2O_3 on metallic mercury in the cold. Mercuric hyponitrite was then prepared by the action of potassium cyanide on mercuric nitrite. This was followed by the preparation of nitrites of alkaline earths in a pure state. It was found that magnesium nitrite was fairly unstable, thus linking it with nitrites of zinc, cadmium, and the alkaline earths. The preparation of the double nitrites of mercury (II) with barium, calcium, and lithium revealed that stability decreased with increasing atomic weight of the metal (P. Ray 1966: 69). This was a key finding in the programme. For details of Ray's later work that is considered relevant see ibid. (69–73).

15. Chatterjee (1986: 30) writes almost despairingly: 'P.C. Ray remains a class by himself. The utter selflessness with which he served the cause of the common people, the spirit of dedication and the Gandhian simplicity of life-style make him a lone and solitary figure. He has left a school and students but the name has all but disappeared.'

4

The Sarton and Coomaraswamy Dialogue

Historians of science have for long been debating the nuances of putting science in context. The project of contextualization of historical narrative has now been turned on the production of historical narrative. Historical narrative too has a context of production. This essay seeks to situate a dialogue between two historians belonging to very different disciplinary genres. There is an underlying epistemological and cultural tension that characterizes these distinct genres. In one sense we have here a prefiguration of the issues involved in the 'science wars', one version of which we presently witness. Alternatively, the history of science found it problematic to integrate the contributions of non-Western cultures. We have here a prefiguration of the discussion on science and multiculturalism, albeit in a different vocabulary. This essay attempts to reconstruct an exchange between the Sri Lankan-born historian of art Ananda Kentish Coomaraswamy and the Belgian born historian of science George Sarton during the decade 1930–40.[1]

A 'charitable interpretation' of the dialogue between the two interlocutors is proposed, but one that is probably more charitable towards Coomaraswamy than Sarton.[2] Sarton was instrumental in establishing a professional identity for the history of science as a discipline within academe (Thackray and Merton 1972). Coomaraswamy on the other hand has been quite marginal to contemporary discussions on the history of ideas. George Sarton, as Thackray and Merton (ibid.) have pointed out, endowed the history of science with a professional identity. However, his role in conferring upon it a cognitive identity is presently considered marginal on account of a lack of currency in the renewal the history of sciences has been subject to in the last two decades. Ananda Coomaraswamy remains an antiquarian figure, mentioned in the popular literature on Eastern mysticism, but better known in the history of art and comparative religion. His writing is strongly inflected with cultural pluralism, while there is an underlying universalism residing in the philological project from which he drew intellectual sustenance. In the exchange between the two we encounter concerns in vogue relating to the science and narrative debate, a debate that has prevailed upon the more conventional concerns of post-Kuhnian philosophy of science. In contrast to the proliferation of disciplinary empires this debate has given rise to in the social studies of science (Galison 1996) both Coomaraswamy and Sarton were classicists seeking a rapprochement across disciplinary boundaries.

This essay does not intend to portray the vocation of their marginality, but discusses two respective objects of investigation, the history of science and the history of art. The distinct addresses of the two interlocutors vis-à-vis their object of investigation reveal very different political predilections. Despite divergent ideological orientations, the two shared a fascination for divulging humanity's common endeavour: Sarton for narrating the history of this endeavour and Coomaraswamy for discovering the dialect of the 'common universe of discourse'. This essay shall also put in context the thematic unity underlying two distinct discursive traditions. The distinction derives from the circumstance that one disciplinary formation drew its inspiration from the Cartesian world-view, while the other was inspired by 19th-century philology. The encounter itself stands out as an exemplar of a very cosmopolitan dialogue.

TWO COGNITIVE DOMAINS

The reconstruction that follows is based on an exchange of letters between Sarton and Coomaraswamy between 1934 and 1947 (Moore and Coomaraswamy 1988). This exchange is coloured not merely by the apparently disparate professional and cognitive orientations of the two interlocutors, but equally by the tension characterizing the romantic narrativization of a domain and its scientistic version.[3] Even though there are few biographies of Sarton, his opus is staple fare with most historians of science. As is well known, he was born in Ghent in Belgium in 1884, and in his younger days was politically inclined to the Fabian brand of socialism rather than to Marxism (Thackray and Merton 1972: 477).

The history of science emerged in the early 20th century as a discipline that sought to chronicle the rise of reason, and the founding role of reason in the rise of scientific thought and the mechanical arts. Europe was seen to have the primary role to play in this historical development (Pyenson 1993). The historian of science Lewis Pyenson (1989) considers Sarton to be one among 'the three men who discovered the history of science during the decade before the First World War', the other two being William Osler and Max Weber. Sarton was probably the most effective evangelist of the history of science during the first three decades of the 20th century, and his conception of the discipline was embedded in the historiography of internal and external history of science. In this scheme, Sarton conceded space for sociologically and philosophically oriented accounts, but indicated that his own passion lay with the internal organization of science (ibid.: 360). The positivist conception of science that had received a minor setback between the wars was to remain with him throughout his life. He spent important years of his life campaigning for a professional identity for the history of science, and was successful in obtaining university positions for the discipline. He founded the journal *Isis* that continues to be an important forum for the publication of articles relating to the history of science (Thackray and Merton 1972: 476).

Sarton shared important elements of the historical method with the other members of the aforementioned trinity. According to this shared vision, history had to do with the movement of people, and the change of ideas and structures through time. The study of the reception and transmission of ideas was to be complemented by 'studies detailing the genesis and evolution of genius' (Pyenson 1989: 367). In this synthetic view of history the comparative method was of central importance, which in turn required a mastery over the classical languages (ibid.: 368). The articles that appeared in *Isis* combined a multitude of perspectives with the historical, such that history could attain its 'full significance'. This would be a step towards reaching the goal of acquiring an 'understanding of the nature of man' (Thackray and Merton 1972: 478). This chronicle of reason was founded on a fundamental European 19th-century trope, according to which the history of science as a history of the progress of humankind could subsume the history of human thought and civilization. This chronicle of reason could re-present the evolution of humanity itself (ibid.: 479). Sarton was to write: 'The history of science is the only history which can illustrate the progress of mankind.

In fact, progress has no definite and unquestionable meaning in other fields than the field of science' (Sarton 1957: 5).

Ananda Kentish Coomaraswamy was born in Colombo, Sri Lanka, in 1877 to a Sri Lankan Tamil father, Sir Muttu Coomaraswamy, and an English lady of a wealthy Kent family, Elizabeth Clay Beely (Lipsey 1977: 7). His biographers have been miserly about the details of his life, respecting his singular aversion to the genre of biography. Lipsey (ibid.: 3) quotes Coomaraswamy on the sort of biographical portrait he respected:

if an ancestral image or tomb effigy is to be set up for reasons bound up with 'ancestral worship', this image has two particularities, (1) it is identified as the image of the deceased as the insignia and costume of his vocation and the inscription of his name, and (2) for the rest, it is an individually indeterminate type, of what is called an ideal likeness .... The whole purpose of life has been that this man should realize himself in this other and essential form.

Respecting his preferred genre of biography, we shall allude to the summary details of his life. His working career can be classified into three major periods. The years 1877 to 1917 were the years of apprenticeship, wherein his vocation was scientific. These years were marked by journeys and a change in profession, at the end of which he settled down to a curatorial position at the Boston Museum of Fine Arts. During these three decades he repeatedly travelled across the globe: India, Sri Lanka, England, and finally the United States. The second phase extends from 1917 to 1929. These years, Lipsey informs us, were years of scholarly work and publication, and the mood was sophomoric for a man moving into middle age. Coomaraswamy reminisced about these years wherein he had 'an agreeably unsettled personal life' (Coomaraswamy quoted in Lipsey 1977: 5). His renowned collection of essays, *The Dance of Shiva* (Coomaraswamy 1918), and the celebrated scholarly tome, *History of Indian and Indonesian Art* (Coomaraswamy 1927) were written during this period. The last phase extending from 1929 to his death in 1947 is the most memorable in terms of his writing on art, religion, metaphysics, and culture (Lipsey 1977: 5).[4] It was during this period, while living in Boston, that his friendship with the 'obstinate historian of science, George Sarton' blossomed (ibid.: 251). Sarton, as editor of *Isis*, was responsible for seeing two of Coomaraswamy's pieces published in the journal (Coomaraswamy 1943/1947; Coomaraswamy 1944/1947).

During the first phase of his life, we learn that Coomaraswamy had graduated with a Bachelor's degree in geology and botany from the University College, London, and was amongst the founders of the Mineralogical Survey of Ceylon. He was the first director of the Survey, and participated in extensive geological expeditions both in Sri Lanka and India. In 1906 London University awarded him a D.Sc. for his researches into Ceylonese mineralogy and his scientific publications on the subject (Lipsey 1977: 12). Two years prior to this he had acquired a place in the annals of late 19th- and early 20th-century geology. Following Mendeleev's path-breaking research on the periodic table, geologists and chemists of the time had set out on a massive hunt for elements and ores (Bensaude-Vincent 1986) that would fill up the gaps in the table. Coomaraswamy, true to the tradition of fieldwork undertaken within the 'high geological tradition', discovered a new mineral, thorianite (Lipsey 1977: 16; Coomaraswamy 1904), an oxide of thorium and uranium.

His training as a geologist left an indelible imprint on his subsequent work, for throughout his life he carefully collected and inventoried art objects with the scrutinizing eye of a geologist examining the archaeology of human artefacts and not the facts of nature. The professional departure can be exemplified through drawings and posters. For example, the tools of the geologist's trade and the artistic conventions of the Northern scientific renaissance appear on the bookplate of his first published book (Lipsey 1977: 16). In 1906 he delivered a number of public lectures on some Kandyan crafts such as steel making, weaving and pottery. The abstract of his talk on weaving reads: 'Different sorts of cotton cloths; the loom and process of weaving. ... Evil nature of aniline dyes; why exploited. Degeneration of textiles crafts in India and Ceylon owing to the demand for cheapness and mechanical workmanship. Indifference of Sinhalese. Cause and decay of popular art in Europe.' The instinct of the historian of art to inventory art objects and their evolution is evident. The latter part of the notice betrays the speaker's and his audience's discontentment concerning the decline of the Sinhalese crafts traditions. Also present are the elements of the late 19th-century Victorian critique of the culture of mass production.[5] It was during the years of field trips in Ceylon that he noticed the decline of the craft traditions as they reeled under the impact of colonial rule. He came around gradually to the view that 'colonialism and industrialism were gross acts of philistinism antithetical to the native genius in art, architecture and music' (Visvanathan 1985: 39). The cover plate of his second book was engraved by Eric Gill in 1920. The symbols of the scientific renaissance were replaced by the symbolism of the East, its sacred art and religion (Lipsey 1977: 119). Lipsey (ibid.: 13) nevertheless conjectures that in the later years of his life Coomaraswamy eschewed sentimentality in discussions on art, and that his unemotional and factual tracts on Hinduism and Buddhism are presented as such on account of his grounding in science.

THE ART OF HISTORY

The two cultures' dichotomy of literature and science is premised upon the epistemological privileging of scientific truth and its modality of unravelling certitudes about the natural world. The ideology of scientism derives from premises that place science at the summit of the pyramid of knowledge; other modalities of knowing and cultural expression are debarred entry into the theatre of progress, or are subordinated to the methodological imperialism of the sciences (Snow 1969). Consequently, the more enlightened studies on the relation between science and art and other forms of cultural enunciation have often donned the hat of what are considered 'influence studies'. These studies assume, either implicitly or explicitly, science as the source of influence, whence science is conceptualized as a 'transcendent rather than a cultural enterprise' (Hayles 1989: 6). Recognizing science as a culturally embodied enterprise entails the recognition that the sciences are embedded within a social matrix and hence constitutes a complex field characterized by manifold social and discursive activities. Lipsey's remark on Coomaraswamy's style, ironically enough, conforms to the essence of the influence model since it traces a 'one-way line from science to literature' (ibid.: 7). In contemporary discussions on science and literature three lines of investigation have emerged. The cultural approach conceptualizes literature and science to be co-produced cultural forms (ibid.: 19). The approach is oriented to a scrutiny of fault lines. The challenging task is to evade radical relativism for it appears to be a natural concomitant of the approach.

The encounter of the histories of two distinctly conceived objects of discourse may be initiated through the reading of Coomaraswamy's metaphysical work, possibly catalogued under comparative religious thought, *Time and Eternity* (hereafter *T&E*). The book was published in 1947—the year of his death—and then republished in 1989. This metaphysical piece may be read by a historian of science as a history of ideas about time and eternity during the ancient and medieval periods (Coomaraswamy 1989). In this reading of *T&E* we find homologies with contemporary notions of textuality, homologies that are analogous in form with some of the core insights of post-structural critical theory, as well as with the social construction of scientific knowledge (Edwards 1994). Despite these differences there is an underlying unity of the goals being pursued by Coomaraswamy and Sarton. Let us examine the structure of *T&E* a little closely to appreciate the levels of the engagement between Eastern and Western knowledge forms.

In Coomaraswamy scholarship *T&E* is considered to be a work on religious exegesis. The book is divided into six chapters, and with the exception of the introductory chapter, the other five deal with theories of time in the cosmology of the great religions: Buddhism, Christianity, Islam, and Hinduism. The exposition of the Greek cosmos is labelled Greece, which is a geographical label rather than a religious one. The discussion on contemporary conceptions of time appears in the chapter on Christianity. It may be conjectured that the Judeo-Christian religions provided the template for the chapterization. The form of Judeo-Christian religion was transposed onto Hinduism and Buddhism. Greece was perceived as the founding source of the modern secular objectification of time. Similarly lines of continuity could be drawn from time as conceived within the Christian traditions and its modern manifestations.

The two central problems of religious exegeses were seen to be the problem of evil, and that of the relationship between divine Omniscience and man's free will (Stoddart in Coomaraswamy 1989: 1). In secular philosophical discourse the mystery of time, that is apparently endless, and the mystery of space, that is seemingly infinite, is of foundational significance. Stoddart conceives of a homology between the religious and philosophical formulations of this central problematic in the manner in which they are posed in Indian philosophy. The mystery of time is ensconced in the principle of change and embodied in the religious imagination in the deific figure, Shiva, an intimidating presence unleashing the forces of destruction. The mystery of space finds expression in a principle of conservation and its iconic embodiment is incarnate in the deific figure, Vishnu, an affable deity representing the non-constrictive side of nature (ibid.).

Coomaraswamy seeks to reconcile Hindu and Greek emanationism with Semitic creationism. The equivalence between Hindu and Greek emanationism is predicated upon an agreement of Vedantic and Platonic ontology, according to which things are not for real, for though they exist, like imitations they are not the real thing

(Coomaraswamy 1989: 3). The reconciliation between these two conceptions is possible through spiritual contemplation; 'the religious preoccupation with life', 'an everyday spontaneism', and 'the preoccupation with the life of experience'. In this interregnum of time that constitutes life, 'the true Christian is really expected to be ... as much as the Sufi, a son of the moment ... as much as the Buddhist Arahant ... for whom there is neither past nor future' (ibid.: 75).

Interestingly, Coomaraswamy does not interpret Vedanta as a doctrine of illusion. On the contrary, he translates the conventional account by a phenomenological one that is clothed in the vocabulary of the historian of arts and manufactures. Coomaraswamy clarifies that the characterization of Vedanta as a doctrine of illusion, that the 'world is the stuff of art' (*maya-maya*) is a misreading. Instead the view distinguishes between the 'relative reality of the artefact from the greater reality of the Artificer (mayin, nirmankara)' (ibid.: 4). The relation between the artefact and Artificer is analogous to the relation between being and Being, for the presence of the artefact persists in the Artificer.

But the issue of relevance to this essay has to do with how Coomaraswamy presents his reading of the different theories of temporality. The historian and philosopher of science would be alerted to aspects of this narrative. The first aspect has to do with the ecumenism of this transcendentalist. For Coomaraswamy there existed a common dialect that bound all human beings into a community, 'homo communis', and this dialect was sacredly ordained. Furthermore, while each civilization and its subcultures could display widely evident differences, there existed a domain where the severalty was subsumed as identity. Coomraswamy drew upon the concept of *bhedabheda* from the Indian school of logic, the Nyaya-Vaisesika, to qualify this notion.[6]

He maintained a distance from a foundationalist project, and this is evident from the absence of any notion of doctrinal purity in any of the chapters. A characteristic feature of the exposition of the notion of temporality in any one tradition is the juxtaposition of similar notions from thinkers in other traditions. And this cross-cultural comparison violates any linear temporal ordering of the appearance of these notions, in which case it was not considered important to establish a lineage or genealogy of ideas or influences. Consequently, it is not a conventional history of ideas, since the underlying premise appears to be that similar ideas surface across very different civilizations at different moments of their history. Furthermore, it is nearly impossible to make any serious claims in the name of originality. The idea was an essential element of the older historiography of science that sought to distinguish between independent invention and simultaneous discovery. In Coomaraswamy's historiography anything new that has been said in the past will be uttered afresh in some future epoch, and the novelty that awaits us in the future is prefigured in the voices of the past.

The idea is developed explicitly in a letter written to Benjamin Farrington (1953), author of the classic *Greek Science*. Coomaraswamy faults Farrington for the view that Greek science and civilization was deeply indebted to the older civilizations of the Near East (ibid.: 13). Coomaraswamy accuses Farrington of methodological presentism: 'what might have been described as physical [in] pre-Socratic thought is really theological thought, since the 'nature' they were trying to explain was not our natura naturata but natura naturans, creatrix universalis, Deus' (*cf.* letter dated 8 October 1945 in Moore and Coomaraswamy 1988: 249). Further, he points out that it is difficult to impute uniqueness to 'any local thought', for there only exists 'local colour'. Methodologically, he clarifies: '*I try never to expound any doctrine from a single source ... I cannot ... conceive of any valid private axioms*' (ibid.: 250, emphasis mine).

This historiographic commitment does not interlock with the sentiment of nationalist historiography. However, nationalism itself acquires different meanings in varying historical contexts, and we must recognize that Coomaraswamy was an ardent supporter of the freedom struggle in India. In any case, history of science framed by nationalist historiography is characterized by the presence of signifiers of pride, and these were manifest within the discourse in priority disputes. The latter are engendered by two sets of factors. First, the measure of advancement of a civilization from the 19th century onwards was its contribution to the progress of scientific and technological knowledge. The pride accruing to the nation was proportional to the continuity of contributions from antiquity into the age of modernity. The gold dust of this esteem certainly rubbed off on the scientists themselves. And thus—this is the second factor—the priority dispute as a source of controversy and investigation marked the preoccupation of scientists with their place in history.

Coomaraswamy concedes within the framework of the history of ideas that in the Sufi, Islamic, and Christian realms the doctrine of 'time and eternity' derived from Platonic–Aristotelian sources, but that by itself was not important. The true value of historical studies resided in the demonstration of the 'universality of fundamental ideas'. This view of historical scholarship was posited to be different from the historical tradition that viewed fundamental ideas as 'the inventions of those who enunciated them' (Coomaraswamy 1989: 66). The priority dispute was unworthy of serious historical consideration, and the invention was far greater than the inventor. Shapin suggests that the projection of the creative scientist-inventor reflected a Western bias that envisioned the act of creation as the outcome of the effort of a unique, individual, and rational being (Shapin 1989: 563).[7] Knowledge, for Coomaraswamy, is produced within communities, much as the artefacts he studied were produced by communities of artisans.

Most organized belief-systems today are tormented at the metatheoretical level by the polarization between objectivist and constructivist distinctions (Anderson 1994: 30). This opposition manifests itself as a 'central cultural opposition of our time', where the objectivist account is committed to the idea of an external, definable, and transcendent authority, while the constructivist account resymbolizes 'historic faiths according to the prevailing assumptions of contemporary life' (ibid.). For Coomaraswamy (1989: 66), the objectivist account, or in his own words, the 'literary history of ideas is of value inasmuch as it is able to answer questions concerning the veracity of a doctrine, or its heretical interpretation'. The objectivist account is subordinate to the transcendental goals of the constructivist one.

THE COMMON UNIVERSE OF DISCOURSE

The exchange between Coomaraswamy and Sarton brims with allusions to the central civilizational tensions of their times, tensions that both interlocutors were committed to surmounting. The two luminaries were situated at two very different posts. Coomaraswamy translates his moves from outside the universalist discourse of science and locates it on a transcendental plane. Sarton's project is located within a notion of science, ideologically moored in positivism.

The correspondence between Coomaraswamy and Sarton reveals that the former found the Western reconstruction of itself, and thus the manner in which it presented its Other, the Orient or Africa, as unsatisfactory, possibly inadequate. The source of this inadequacy is traceable to the connotation of the term civilization (*cf.* letter dated 7 October 1943, in Moore and Coomaraswamy 1988: 13). The term was invariably deployed in the singular, and Greece was projected as the source of civilization. In a letter to Sarton he discusses his differences with Werner Jaeger's work *Paidea* (*cf.* letter dated 7 July 1942, ibid.: 171).[8] This notion of civilization was hegemonic enough to taint the Orient's recuperation of itself (*cf.* letter dated 13 August 1939 ibid.: 41). Consequently, Radhakrishnan's *Indian Philosophy* is labelled a 'Western interpretation of Hinduism' since it had entirely ignored Islam that had significantly mediated between Eastern and Western schools of thought (ibid.).[9]

Coomaraswamy reconceptualizes the term civilization by disassociating it from the dominant Hellenophillia, and projects in its place another exemplar of civilizations. On 22 August 1947, five days after India had declared independence from British rule, he delivered a Farewell Address at Harvard Club, Boston, on his retirement. He summarized his life's work on Indian art, a subject that took him to the traditional theory of art, and the relation of man to his products, and to problems in iconography. The latter drew him to the sphere of comparative religion and metaphysics (Coomaraswamy in Moore and Coomaraswamy 1988: 443). His definition of civilization echoes the Victorian disenchantment with industrialization. His involvement with the *swadeshi* school of Bengal during the first two decades of the 20th century is reflected here. This 'Blakean protest' (Lipsey 1977: 105) finds a voice in a piece called 'Love and Art' published in 1915:

If the advocates of compulsory education were serious, and by education meant education, they would be well aware that the first result of any real education would be to rear a race who would refuse point-blank the greater part of the activities offered by present day civilised existence ... that life under modern Western culture is not worth living, except for those who are strong enough and well equipped to wage a perpetual guerrilla warfare against all the purposes and ideals of that civilization with a view to its utter transformation. (Coomaraswamy 1915, quoted in Lipsey 1977: 105).

Underlying the disenchantment with modern Western culture was the desire to recover the romantic, artisanal, pastoral past that turns away from the hegemonic rationality of a science that was considered iconoclastic enough to demolish the sacred order. In Coomaraswamy's vision of civilization: 'A society can only be considered truly civilised when it is possible for every man to earn his living by the very work he would rather be doing than anything else in this world—a condition that has only been obtained in the social orders integrated on the basis of vocation, svadharma' (Moore and Coomaraswamy 1988: 444).[10] Unfortunately, this Utopia was an apology for a caste-based society; and more than apology it sought to accord these formations legitimacy.

However, this definition of civilization was not exclusivist for it was capable of embracing many more cultures within the civilizational fold. The *bhedabhedavadin* was from this vantage point offering a perspective of human culture that took cognizance of their 'apparent diversity', and in these variations recognized 'dialects of one and the same language of the spirit', for transcending the diversity of tongues was a 'common universe of discourse' (Moore and Coomaraswamy 1988: 444). We shall have more to say about this in the subsequent discussion on Eastern wisdom and Western knowledge.

Orientalist representations of the East, as well as late 19th-century obsessions with the manner in which modern science was presented as a break with the past, accentuated the distinction between West and East. Coomaraswamy, on the contrary, suggested that the idea of East and West as a cultural distinction as different from a geographical antithesis was a post-renaissance invention. This cultural schism was one that 'presents itself only accidentally in terms of geography; it is one of the times much more than of places' (Coomaraswamy 1947: 66). This was tantamount to superposing a cultural or cognitive rupture on a geographical segmentation of the world. The variations in culture were for Coomaraswamy analogous to that of dialects. The comparative method led him to conclude that there 'is a universal intelligible language, not only verbal but also visual, of the fundamental ideas on which the different civilizations have been founded' (ibid.). This axiology posited a 'common universe of discourse' that constituted the bases for communication, understanding and agreement (ibid.).

Evident here is the influence of late 19th-century philology. Two fundamental elements of this project were the comparative method as expounded by August Schleicher in the mid-19th century. The fundamental premise was that languages A and B could be genetically related even if they had no cognates in common. The second could be considered the founding postulate of the philological project, namely, the idea of the monogenesis of language.

Sarton's catholic ethic compelled him to allow Coomaraswamy to articulate his views on the East–West divide. The divide within the circle of historians of science, which the Orientalists had a major role in shaping, were framed within the diptych (Elzinga and Jamison 1981: 9).

TABLE 4.1
The Dichotomy of Eastern and Western Science

East	West
Experiential	Experimental
Aesthetic	Rational
Intuitive	Theoretic

Coomaraswamy's views can be constructed from his review of the work of the French Orientalist René Guénon. The review was published under the heading 'Eastern Wisdom Western Knowledge' (Coomaraswamy 1943). However, while Sarton published the piece, he introduced the article with an editor's prefatory note, or warning. The cautionary note read:

The author of this essay, deeply versed in Eastern as well as Western lore, is the leading mystical philosopher in this country and the most able to study Guénon's views from the inside. The Editor of *Isis* and the majority of its readers do not share those views but welcome an authoritative and sympathetic explanation of them (Sarton quoted in Lipsey 1977: 171).

A positivist readership is put on guard for the author of the essay is projected as a mystical philosopher, rightfully an authority in his domain, but culturally and epistemologically an outsider.

COOMARASWAMY AND GUÉNON'S ORIENT

We shall briefly situate Guénon within French Indology. René Guénon (1886–1951) emerged as a French Indologist of repute between the two World Wars. Intellectually, he is positioned at the intersection of two crises, the one academic and the other denominational (Lardinois 1994: 6–7). Academic Indology in the second half of the 19th century was torn between philologists translating and editing classical texts, and the specialists. The former paved the 'royal path' to the knowledge of the Indian world, and all other disciplines were implicitly or explicitly ordered in a hierarchy subordinate to philology (ibid.: 4). In denominational terms Guénon belonged to the 19th century esoteric movements, characterized as neo-Thomist anti-science thought, that reacted adversely to the new religious sciences and were positively hostile to democratic and egalitarian values that were the legacy of the French revolution. The romantic reaction found a voice in Charles Maurras' right wing party l'Action Française (ibid.: 10). The Vatican affected a schism between Catholic intellectuals and the party, which led to the expulsion of Guénon, for whom in any case metaphysical principles prevailed over the 'contingent aspects of political action' (Ibid.: 10).[11]

Both Coomaraswamy and Guénon were critiques of modernity, for they recognized in the rise of modernity the vitiation of traditional social universes. Consequently, in Guénon Coomaraswamy recognized a kindred soul who saw through the demonization of Western civilization despite a century of progress (Coomaraswamy 1943: 57). What are the elements of Guénon's critique of modernity that Coomaraswamy highlights? For these could well be seen as the signs of his own inquietude. Guénon rejects two cultural tenets of modernity. The first relates to the rhetorical promise of science—the idea that science would usher in the millennium. This rejection of Baconian optimism was coupled with the idea that the success of the modern West was epitomized by a culture based on reason and materialism (ibid.). Guénon rejected both these propositions. The critique was warranted in a society that had been dehumanized by the culture of industrialism, where the fruits of scientific knowledge, that had radically altered social relations, could be abused. It is around this circumstance that doomsday prophecies were framed, for now Western man, having betrayed his sacrament with God, would be enslaved by the logos of science (ibid.) Coomaraswamy, at the end of this account, of which I have offered a very prosaic précis, feels that those seduced by the trappings of modernity would find Guénon preposterous, for his oeuvre lays stress 'on these things, because it is only to those who feel this frustration, and not those who still believe in progress, that Guénon addresses himself' (ibid.).

We may conjecture that Coomaraswamy was not merely heralding the views of an apostate in a sober journal of the history of science and ideas, but through a clandestine act of substitution, Guénon's voice is veritably the voice of Coomaraswamy. Why do I say this? The piece on Guénon just discussed must be read alongside another piece by Coomaraswamy (1947) republished in *Am I My Brother's Keeper*, entitled 'East and West'. The essay affirms in greater detail the principle disagreements and conflicts that are raised in the essay on Guénon, but the latter's name appears nowhere. On the contrary, there is an echo of the debates that occurred in the National Council of Education, Calcutta, during the first decades of the 20th century. This was a movement he was closely associated with during the politically tormentous years of *swadeshi* (Mitter 1994: 260). The debate within the National Council of Education had to do with legitimating a programme on scientific education within the framework of the nationalist struggle. An important element within this `narrative of freedom' related to the course of industrialization that was appropriate to India's skills and resource endowments (Raina and Habib 1993).[12] Coomaraswamy, however, fell out with the swadeshi movement because he felt that the movement's commercial thrust had subverted traditional aesthetics embodied in the crafts tradition (Visvanathan 1985: 39).

Returning to the article on Guénon, Coomaraswamy insists that the East–West distinction was a post-renaissance invention. Further, that Guénon was not an Orientalist. The reason he offers has to do with the central position accorded to pure metaphysics over other forms of knowledge in Guénon's writing. Consequently, rather than portraying Guénon as a French Orientalist, Coomaraswamy portrays him as a descendant of a 'predominantly Eastern tradition' (Coomaraswamy 1943: 58). What argument does Coomaraswamy offer to divest Guénon of the label Orientalist?[13] This will provide us a glimpse of Coomaraswamy's critique of Western civilization. Guénon, his interpreter suggests, advocates a turn to the East, for the East was metaphysical. The West had eschewed the metaphysical in its passage to the age of modernity (ibid.: 59). The metaphysical world was singular, and thereby it could not be clothed in the cultural apparel of an Eastern or Western metaphysics (ibid.). The intent of this turning eastward was not to 'Orientalize the West', for were it so, the project would be no different from the mission civilizatrice, whose 'proselytizing fury' was directed at the distribution of modern civilization (ibid.: 60). Since the metaphysical was the common inheritance of mankind, the encounter with the East would restore to the West those values that had undergone the most sinister transvaluation (ibid.: 59).

This transvaluation was a consequence of modernization. In order to renew itself the West would have to reconstitute the objects of knowledge. These objects could not be the 'facts of science', or premised upon the desire for 'acquiring the power to conquer nature'. Coomaraswamy then introduces a typical late 19th-century Oriental dualism, of an East that excelled in the knowledge of the inner self and the West that had accomplished a degree of understanding of the workings of the material world (Coomaraswamy 1947: 66). Underlying this dualism was the romantic resistance to entertain the law of progress as one of social evolution. Coomaraswamy was to write that both East and West were at cross purposes because the West is 'resolved and economically determined to keep on going it knows not where, and calls this rudderless voyage progress' (ibid.: 66). The critique of progress is relevant, for despite the attempt to paint progress with the ideological neutrality of science, it was instantaneously perceived in this circle as a politically loaded construct. Progressivism and its critique provide us with the two sides of an evangelical programme. This is schematized in the table below.

TABLE 4.2

The Orient and Occident as Seen through the Prisms of Orientalism and Scientism

	Orient	Occident
Orientalism	Spiritually developed	To be infused with Eastern spiritualism
Scientism	To be infused with the science and technology from the West	Scientifically and technologically developed

The purveyors of progress, the torch-bearers of scientism, were committed to the mission civilizatrice. The romantic Orientalists, those expressing their dismay about the vagaries of contemporary Western civilization, pleaded for the spiritual revival of the West, the recovery of its sense of the spiritual. Coomaraswamy's disenchantment with scientific humanism and industrialism had to do with the reduction of cultures to the 'lowest common denominator' (Coomaraswamy 1944: 72). Hence any kind of globalization would be construed as a levelling rather than an elevating force.

THE POLITICS AND OBJECT OF HISTORY

The history of the colonial era is particularly fascinating for the constructedness of historical and political categories is most evident at this juncture. Coomaraswamy's scholarly writings from the last two phases of his life betray a political revivalism, and yet, in the non-political sphere he appears as a figure fairly sensitive and appreciative of cultural diversity the world over. This kind of conservativism interposes itself more obliquely even in Sarton, whose universalism and Fabian socialist background did not suppress his 'latent racism', and 'unreasonable allegiance to the nation-state' (Pyenson 1989: 378). The more important question is that both Coomaraswamy and Sarton shared this desire to reach out to the common universe of discourse.

As Coomaraswamy (1947: 74) put it: 'We need mediators to whom the common universe of discourse is still a reality.' One of the reasons why Sarton rarely figures in contemporary discussion in the social studies of science is that the progressivist faith on which the Sartonian project was founded stands challenged (Thackray and Merton 1972: 480).

Let us briefly recapitulate Sarton's (1957: 5) notion of science:

Definition: Science is systematized positive knowledge, or what has been taken as such at different ages and different places. Theorem: The acquisition and systematization of positive knowledge are the only human activities that are truly cumulative and progressive. Corollary: The history of science is the only history which can illustrate the progress of mankind.

This fascination with global history was the product of the belief in the 'unity of knowledge, in the integrity of experience, and in the need for a holistic philosophy that embraced art and science' (Thackray and Merton 1972: 487). Coomaraswamy's friendship with Sarton was testimony to this project.

Coomaraswamy appears as a figure who recognized that professedly universalist discourses came with political entanglements. It appears that he was not very successful in navigating past them. In any case, his critique of progress did not repudiate the internationalist ideal. He was to reaffirm this ideal, despite the political revanchism that haunted him. 'I want to serve not merely India, but humanity, and to be as absolutely universal as possible—like the Avalokitesvara' (Coomaraswamy quoted in Mitter 1994: 260). The *swadeshi* movement was for him a cultural alternative to colonialism. But as Mitter points out, it was Gandhi, steeped in the political realities of India, who translated this into a political programme. Coomaraswamy's writing displays the symptoms of the migrant intellect, which, as Rushdie writes, roots itself in itself and its 'capacity for imagining and reimagining the world' (Rushdie 1991: 280).

At the end of his professional career he was to suggest that the encounter between East and West had produced two outcomes in India. Two personality profiles had apparently emerged. The one epitomized in figures like Nehru, hagiographers of science and modernity, who were 'queer mixtures of East and West, out of place everywhere and at home nowhere'. The other outcome, in a possible allusion to Gandhi, involved 'being oneself ... in place anywhere ... at home everywhere ... a citizen of the world' (Coomaraswamy 1947: 74–5). This was a reflection of the tension produced within him of clashing self-images, and it is likely that he was himself in the Nehruvian mould, while the self 'at home everywhere ' remained an aspiration. On the contrary, Sarton may have aptly conformed to the latter image.

Hence on his retirement he announced to a Boston audience that having spent half his life as 'a student of the manufactures' (Moore and Coomaraswamy 1988: 443) at Boston, he was returning to India. This return he called an *'astam gamana'*, a going home (ibid.: 444). This aspect of Coomaraswamy, who had spent three-fourths of his life away from either Sri Lanka or India, drawn poignantly to an imaginary homeland called India is interesting.[14] If the West in its encounter with the East were to find itself, then the East would no longer remain the 'mysterious East', an obligatory discursive Other. In case it did not, the world would be reduced to 'the present State of Europe' (ibid.: 475).

The last phase requires a little elaboration. Sarton's passionate quest for the unity of human knowledge was impelled by the moral failure prompted by the World War I (Thackray and Merton 1972: 487). At the end of World War II he was to write: 'In the better kind of world, which we all hope will be fruit of this war, the children will be expected to learn of the evolution of mankind ... . The history of science will teach men to be truthful, it will teach them to be brothers and help one another' (Sarton quoted in Pyenson 1989: 368). Coomaraswamy's search for the common universe of discourse and his anxiety concerning 'the present state of Europe' must be situated within this very circumstance. The foregoing account has been lop-sided in its treatment of Coomaraswamy than would give comfort to the author of this essay. But this has been done to ensure that the historical context within which Coomaraswamy is located is read symmetrically. However, distinct disciplinary commitments and political orientations did not impede the two interlocutors from sharing analogous goals that had acquired greater urgency during the troubled times they lived through.

The historical methods employed while approaching their respective objects of investigation diverged sufficiently. This had to do, among other factors, with distinct perceptions of agency, their divergent experiences as political subjects, and the manner in which they responded to and shaped the dominant ideas of their time: progress, industrialization, the civilizing mission. Despite his radical political predilection, Sarton's positivist inclination obscured his ability to anticipate the passing away of positivism. Coomaraswamy was drawn throughout his life to the Indian nationalist struggle, but even there he found himself at odds with those pursuing the path of critical modernization. However, he remained ecumenical in his intellectual enterprise. His years at Boston were years of engagement with the received project of the history of ideas. Consequently, this student of the arts and manufactures envisioned a theory of knowledge that drew part of its inspiration from the former. Respecting Coomaraswamy's vision of history, which means dispensing with the ideas of individual genius and that of anticipation or priority, it may be conjectured that his unusual location as a colonial subject enabled the articulation of this theory.

THE ELUSIVE RAPPROCHEMENT

This essay has chronicled the situatedness of the dialogue between Coomaraswamy and Sarton across two disciplinary formations, possibly two distinct civilizational projects. How are the tensions that segregate these realms reconfigured? From that iconic milestone in the history of science, the scientific renaissance to the age of modernity, the formations referred to as art and science have moved apart. Contemporary recognition of any reconciliation is seen to be a consequence of the cognizance academic discourse has taken of contributions non-Western cultures have made to the arts and sciences, wherein these cultures are conceived as broadly as possible and not merely in the 'geographical sense of the term' (Massini 1994: 45). In one of the last papers authored before his death, the philosopher of science Paul Feyerabend proposed three theses concerning artistic production. One of these theses underplays the importance accorded to individual creativity and invention: 'If art works are natural products, then, like nature, they will change, new forms will appear, but without major contributions from isolated and creative individuals' (Feyerabend 1994: 89). Furthermore, even in the sciences and metaphysics, invention cannot be ascribed to the act of genius of a solitary thinker. Novelty is a feature of natural processes (ibid.: 91). The arts and the sciences do not embody any dichotomy in nature, and are temporary categorizations, for both scientists and artists learn by creating artefacts (ibid.: 93). In making this claim Feyerabend is not being a relativist as he is often made out to be. He clarified in one of his last books that ultra-relativism precluded the possibility of learning new ways of life. As long as the possibility of understanding existed any system was potentially all systems, and any claim made relative to a given system lost its 'power as a general characterization of knowledge' (Feyerabend 1991: 152). Thus, the historian may encounter incommensurability between the belief systems of varying historical situations and actors, or between the researcher and the historical actors being researched. However, Kuhn and Feyerabend argued that incommensurability could be surmounted either by becoming bilingual, as suggested by Kuhn, or by enriching one's language, as suggested by Feyerabend (Biagioli 1996: 191). The Coomaraswamy–Sarton encounter may be seen as one such attempt to bridge the gap between the two worlds by enriching each other's understanding through such cross-talk.

This essay is itself occasioned by a renewal in the social studies of science that seeks to integrate discourses long considered marginal to the master narrative of the history of science into a larger picture. We can find parallels and filiations between Coomaraswamy's cross-civilizational hyper-textuality and the contemporary reconfiguration of the object called science. As the techno-scientific regime installs itself more securely within our life worlds, we witness a greater disunity in the images and representations of science.

Both Coomaraswamy and Sarton lived the mature years of their lives in dark and trying times. Sarton's disciplinary response was the quest for the unity of mankind in humankind's scientific production and learning. Coomaraswamy, when confronted with a universalism he found problematic, was compelled to search for a common universe of discourse that would be divested of the hegemonic contamination of the times. This may have been a radical intellectual undertaking, but Coomaraswamy was all along fairly conservative. In our own times we could ask whether postmodernism would have as radical an appeal for third-world societies. One view suggests that postmodernity is contiguous with the modes of thought that formed the bedrock of colonialism, that it is the third stage in the occupation of the non-West. And that it had the potential of becoming the master alibi for the 'continued exploitation and oppression of non-Western cultures' (Sardar 1993: 882). In a postmodern world Coomaraswamy the *bhedabhedavadin* and Sarton the positivist historian would still have been in dialogue seeking out the unity of humankind.

NOTES

1. Eric Hobsbawm (1993: 6–8) refers to the period between World War I and II as the 'Age of Catastrophe'. During this interregnum 'Western

civilization' of the 19th century witnessed a breakdown, and the climate of political liberalism went into retreat.

2. For a discussion on charitable interpretation see Steven Shapin and Simon Schaffer (1989).

3. Coomaraswamy and Sarton may not have in their individual capacities subscribed to the terms of the divide, but each discipline is defined by a collegial circle that is bound together by literary technologies (Shapin and Schaffer 1985).

4. The astrophysicist S. Chandrasekhar (1991: 48) points out the different manifestations of creativity in the life of a scientist and an artist: 'In 1817, at the age of forty-seven when the long period of meditation, during which Beethoven composed very little, was coming to an end, he said to Cipriani Potter with transparent sincerity: "Now, I know how to compose." I do not believe that there has been any scientist, past forty, who could have said, "Now I know how to do research." And this to my mind is the centre and core of the difference: the inability of a scientist to continually grow and mature.'

5. For a study of the political milieu in which Coomaraswamy found himself, and the influences on him see (Mitter 1994).

6. The concept of *bhedabheda* was proposed by the Nyayika Sridhara in the 10th century, in his work the *Nyayakandali*, which was a commentary on Prasastapada's *Padarthasamgraha*. The concept is translated as 'identity-in-difference' in English. As Sridhara writes: `Some people argue as follows about universals: the universal is identical with its instances. We do not have judgments about two distinct entities, as in seeing a man with a stick, and we do not see a cow as qualified by a distinct entity cowness. ... Each individual thing individuates itself and like-wise classifies itself as of a certain kind. This explains the view known as bhedabheda or identity-in-difference. A universal is identical with each of its instances, which are different from each other. This being what is found to be the case, it is pointless to complain that a thing cannot be both the same with and different from another thing at the same time: this is just how things are' (quoted in Potter 1977: 519). The quotation is based on Durgadhara Jha Sarma's edition of the *Nyayakandali* translated by Ganganatha Jha.

7. This reconceptualization within the sociology of science is echoed in Latour (1988: 14): 'When we are dealing with scientists, we still admire the great genius and virtue of one man and too rarely suspect the importance of the forces that made him great. ... The great man alone is alone in his laboratory, alone with his concepts, and he revolutionizes the society around him with the power of his mind alone. Why is it so difficult to gain acceptance, in the case of great men of science for what is taken as self-evident in the case of statesmen?'

8. Werner Jaeger was a professor at Harvard at the time.

9.   Ironically enough, a later-day interpretation brings Radhakrishnan and Coomaraswamy on one side, and Kosambi and Debiprasad Chattopadhyaya on the other, as two attempts to 'reconstruct Indian culture according to categories which would seem internally consistent to the Western mind' (Nandy 1983: 82).

10. The first part of this vision of a civilized society shares features exemplified in Marx's vision of a classless society. But the comparison stops there, for the latter half of this vision accords legitimacy to caste-based society.

11. Coomaraswamy (1918) was sympathetic to the Indian nationalist cause, but there were many Indians within the struggle who found him an 'uncomprising reactionary'. Cf. the essay on 'Young India' appearing in the collection of essays *The Dance of Shiva*. It is interesting to note that the French historian of science Pierre Duhem was a devout catholic and a French nationalist (Cohen 1994: 49). Despite Guénon's and Duhem's distinct epistemic commitments they shared the same political leanings.

12. Lyotard (1984: 32) suggests that the 'narrative of freedom' is one of the narratives of legitimation resorted to when the state 'assumes direct control over the training of the people, under the name of the nation, in order to point them down the path of progress'.

13. Lardinois (1997) points out that Guénon was a philosopher of doctrinal esoterism seeking to unveil tradition. This involved the task of revealing the immutable and eternal principles that assisted in forging the relationship between men and the socio-cosmic world. Underlying this project were two postulates. First, that Hinduism was constituted as a metaphysical system that could be apprehended along metaphysical lines alone. Second, Guénon insisted on the superiority of the indigenous point of view. For the 'indianiste' the Brahmins were the 'authorized interpreters' of this tradition and it sufficed for the former to be their spokesmen in order to be able to express the truths of this tradition (ibid.: 12–13).

14. Rushdie (1991: 12), as a contemporary immigrant writer, remarks: 'the past is a country from which we have emigrated ... its loss is part of our common humanity ... the writer who is out-of-country and even-out-of-language may experience his loss in an intensified form'.

5

Science, Scientists, and the History of Science in India (1966–94)

This chapter reviews a genre of the history of science appearing in the *Indian Journal of History of Science* that strictly observes the internal–external divide. The attempt here is to identify the historiographic elements that constitute this genre of the history of science in India. One of the elements of this review is a bibliometric analysis that would enable the identification of the priorities of historians of science in India publishing in the journal. These priorities and the underlying historiography render certain kinds of problems amenable for research and investigation, and foreclose the pursuit of others. In attempting a social epistemology of the discipline (a very preliminary one is proposed here), other themes and areas may be identified. In addition, it is suggested that the conservatism of historians of science in India has curtailed the growth of the discipline. While this review is partisan, it also seeks to evaluate the growth of the discipline in terms of the precepts set down by the founders of the journal itself.

The *Indian Journal of History of Science* (hereafter *IJHS*) is a journal brought out by a scientific society, the Indian National Science Academy, and publishes papers solely in the discipline of the history of science and technology. As will be discussed, the papers appearing in the *IJHS* fall within a certain genre of the history of science writing. At the outset this is simultaneously an exaggerated claim, as much as it is trivial for the sociologists of scientific knowledge. In any case, it may be reasonable to suggest that most of the papers published in the journal are authored by Indians undertaking research in the area, and that a significant proportion of Indians researching the history of science publish in the *IJHS*. Hence, a simple bibliometric analysis of these publications would offer us some insight into the sociology of the discipline.

A feature that both the natural and social sciences have globally witnessed is a proliferation of journals. While this is true of the domain referred to as STS (science, technology, society studies), the *IJHS* has till recently remained the sole academic forum for the history of science in India (several other short-lived efforts notwithstanding).[1] However, those working on the history of sciences as it pertains to the geographical or cultural boundaries of India have, with a shift in perspective, analytical, or historiographic focus, been publishing in other journals as well. The dispersal of publications into a multitude of journals is surely not a random phenomenon, since journals are today (as was true in the past as well) associated with research communities and networks, tethered by methodological concerns, theories of interpretation, a domain of investigation, and within the domain a shared perception of the essential problematic.

A cursory examination of the bibliometric trends would provide us with the broad contours of the discipline in India over the last 30 years. From quantitative data it should be possible to infer the rate of growth of the discipline, the comparative growth of sub-disciplines, the proportions by which the community of researchers is expanding, the degree of specialization in each of the sub-disciplines, the key themes of research, and the core debates that cropped up in the domain. On the qualitative front it would be meaningful to draw upon some inferences regarding the principal historiographic frames adopted by those publishing in the *IJHS*, and to distinguish the different temporal phases in this historiography.

As has been indicated elsewhere, the history of science in India as a project situated within the discourse of modernity dates back to the efforts of the British, French, and German Orientalists, who sought to chronicle India's cultural and scientific past, on occasion to unearth gems that had been lost in the labyrinths of historical consciousness (Filliozat 1964), and on occasion to legitimate British dominion over the country (Nandy 1988).[2] In either case, the vantage point appeared to have been the evolving frontier of modern science, and the mechanistic-empirical frame was the one within which 19th-century philosophy of science was ensconced. Science in 19th-century Europe had just emerged from the great theoretical consolidation of the 18th century (Gascoigne 1996), where disciplines like physics and chemistry emerged out of natural philosophy, and disciplines like geology and zoology speciated from natural history. In the light of the exponential growth of the sciences since then, a great deal of the history of science written in the 19th century would require an appreciation of no more than science taught at the undergraduate level today. On the historiographic front nevertheless we have a buzzing hornet's nest to contend with.

The Orientalists provided us with the first histories of science or traditional knowledge relating to the natural world as it obtained in ancient and medieval India. In historiographic terms, Orientalism as a Western discourse has over the last couple of decades come in for voluminous and scathing criticism (Said 1978). This substantial critique of Orientalism was itself the product of the renewed gaze instituted by the newly independent nations—the former colonies (Said 1994). Amongst the many defects of Orientalism as a discourse was its ability to encompass a great deal of social and cultural diversity within far too broad categories—put differently, it painted the entire Orient with a very broad brush. This is not an error we should replicate in contextualizing Orientalism and the Orientalist archive as homogeneous.

While being predisposed to the critique of Orientalism, we concurrently require a nuanced appreciation of the Orientalist representation of the Levant and the Orientalist representation of India; and there are subtle distinctions to be drawn between the British, French, and German Indological traditions (Thapar 1993A).[3] The peculiarities in these traditions could, among other factors, be ascribed to the different political posts occupied by the Indologists belonging to these traditions. However, the tracts of the first generation of Indians writing about the history of science indicate that Orientalism initially shaped the emergence of the tradition of history of science in India. As the struggle for independence grew in extent and intensity, this tradition, inaugurated by P.C. Ray, departed from certain axioms of Orientalist historiography.[4] This departure is highlighted in the corpus of Acharya Prafulla Chandra Ray's *History of Hindu Chemistry*, and further in a large number of popular histories and lectures delivered by him.

While the *IJHS* embodies one representation of the science–society relationship as construed in Indian scholarship, it is a matter of immediate interest that one of the peripheral outcomes of the professionalization of science in modern India is the attempt to

professionalize the history of science. However, this professionalization as manifest in a distinct cognitive or institutional identity for
the discipline is not evident. The practice of the history of science at
the end of 100 years continues to be afflicted with the drawbacks of
'dotage theory' (Russel 1988).

Hence, rather than examining the avenues of professionalization
of the discipline, it is more relevant to explore the images of science
that pervade the community of Indian scientists. Images are not
directly mappable onto the realm of the 'real', but nevertheless shape
responses and actions. Belonging to Popper's World 3, there is a
sense in which they are rendered 'real'. This essay proposes that the
view of science that distinguishes the Indian scientific community re-
emerges in the *IJHS* account of the history of sciences in India. This
image of science in turn legitimates the status of the scientist as the
generator of truths about the natural world. The scientific community draws capital from the resource base so created and social status
obtained.

The decade starting 1950 was witness to a rapid growth of the
institutions of scientific research in the country. History of science
studies in India date back to the second half of the 19th century, but
the discipline begins to be organized as one in the 1960s. In 1959 the
National Institute of Sciences of India (NISI)—now the Indian National Science Academy—constituted a board that would address
the task of producing an authoritative history of sciences of India.
The absence of a concrete organizational charter for the history of
science and the dearth of resources hampered the development of the
field for some time. Five years later, in 1964, at the behest of the
Ministry for Scientific Research and Cultural Affairs, a meeting of
interested scholars was organized, where it was decided that the
NISI would take on the responsibility of constituting a National
Commission for the Compilation of History of Sciences along lines of
the UNESCO Commission for Mankind.

With the passage of time, the National Commission delegated the
disbursal of funds for research for the history of science to the NISI.
It was suggested that these funds be utilized for creating professorial
posts and a cadre of research scholars in the discipline. As was true
of any other discipline, it was hoped that this intervention from
above would catalyse the institutionalization of the field, and it was
desired that university departments recruit faculty into departments
of history of science. These faculty members would be drawn from
the reservoir of scholars trained by the experts appointed by the

National Commission. However, after three decades following the enunciation of this programme, the institutionalization of the history of science in university departments has still to take place. Further, the original charter of the National Commission recommended that researches initiated by the former would be published in the form of bibliographies, monographs, and research articles in the *Indian Journal of History of Science*. The proposal was approved in 1965.

What were the key concerns of the historians of science in India who published their work in the *IJHS*? In order to answer this question it is necessary to raise a closely associated set of issues having to do with the mirroring of the perceptions of scientists and their understanding of science in historians' accounts of science. Having done so, a key proposition underlying this essay would stand elucidated, the proposition being that the absence of professionalization in the history of science, amongst other factors, arises from its dependent existence. In other words, the *IJHS* history of science provides the 'soft' garnishing for Indian science, and tends to serve as an apology for science in India, both past and present.

Furthermore, from the perspective of the historiography of science, it may be suggested that a sizable number of publications appearing in the *IJHS* are situated within a historiographic framework that clearly demarcates the internalist from the externalist account of science.[5] In addition, an occasional perusal of the journal may give rise to the impression that it is devoted to the internalist account of the history of science. If that be the case, then the philosopher of science would not be wrong in asking the larger question that if the internalist tradition of the history of science is so strong in India why have we been unable to propose an epistemological approach to the history of science. At this stage it is suggested that the object called science is taken as a positivist given. At the other end of the internal–external divide a critique often voiced is that a sociology of science situated in India's historical experience is yet to emerge. These limitations taken together, it may be suggested, have prevented the emergence of a comprehensive picture (the 'Big Picture') of science (Cunningham and Williams 1993), or of a synthetic theory that discloses the nature of science in the Indian context.

THE BIBLIOMETRIC BACKGROUND

The *IJHS* commenced publication in 1966, and between then and 1994 about 568 research papers, articles, and translations of texts

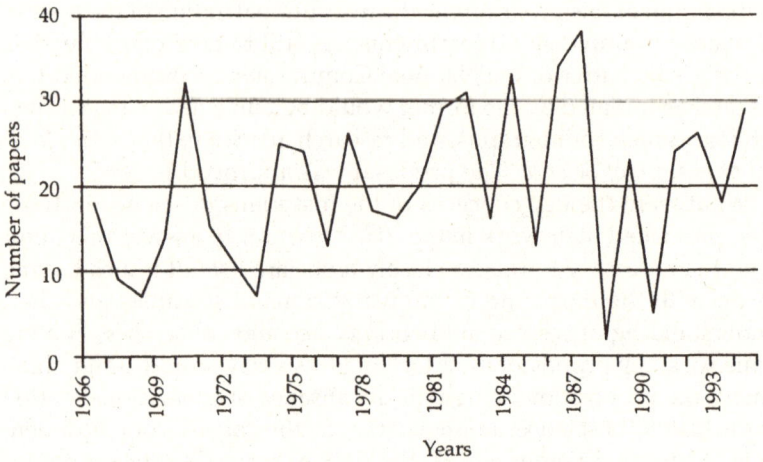

Figure 5.1: Number of Papers Published Annually in IJHS

have been published. The number of papers published annually increased from 18 to 29 (see Figure 5.1): this exhibits an annual growth rate of 2.17 per cent. But this is not a representative figure, for the maximum number of papers published in any year was in 1986 and 1987—34 and 38 respectively. The mean number of papers published annually over this time period is 20.32. The increase is thus marginal, which is borne out by the fact that 18 papers appeared in 1993. In over 28 years the discipline has not grown appreciably if we were to use the number of papers published annually as an indicator.

Consider the number of papers published for each of the sub-disciplines (see Figure 5.2), such as the history of astronomy and mathematics, the history of medicine, the history of the life sciences, the history of physics, chemistry, and alchemy. Between 1966 and 1994 about 36 per cent (204 papers) of the papers published have to do with the history of astronomy and mathematics. In the ancient and medieval periods astronomy and mathematics are taken together. A subject of current importance such as the history of technology accounts for a mere 5 per cent of the papers. The rest includes the histories of physics, alchemy, geology, mineralogy, and agriculture, while science refers to papers reflecting the Weltanschauung of the sciences. There are 54 research papers on the histories of physics, chemistry, and alchemy and 12 in the area of agricultural science. The history of medicine and life sciences accounts for the largest number

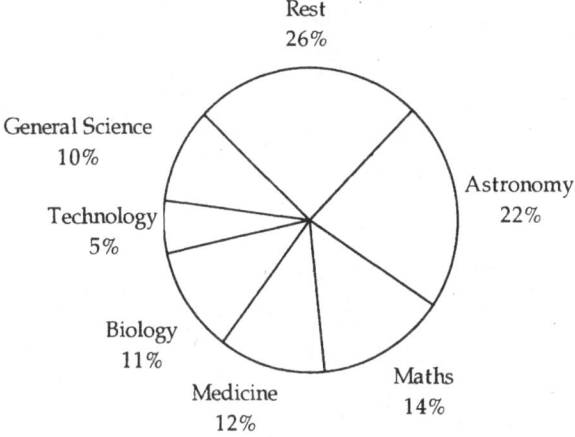

Figure 5.2: Distribution of Papers Published in IJHS

of papers after the history of astronomy and mathematics—131 papers (16 per cent). This indicates that the community of scholars publishing in the *IJHS* were predominantly disposed to the study of the exact sciences and the history of medicine.

In the next instance, consider the publications in the sub-disciplines and the corresponding historical periods that are the subjects of the papers. The papers have been classified into periods relating to the ancient, medieval, late medieval, colonial, and contemporary periods (see Figure 5.3). More than half the number of papers written are situated in the ancient and early medieval periods (52 per cent). The modern and contemporary periods (> 1900) account for about 21 per cent of the papers. There is then a very light middle spanning 1000 years of history, that is covered by 27 per cent of the total number of papers published.

It is natural that in the first phase of the history of science and technology in India the focus is the ancient period, since clarity on this front would provide intelligibility on concerns relating to foundations. However, if this continues to be a dominant concern of the papers published in the journal, then one of two things could be inferred: either members researching other aspects of the discipline are publishing in other journals or that little research is being pursued on science during other periods of Indian history.

The figures suggest that since the 1980s a larger proportion of papers are dedicated to the history of science of the medieval, late

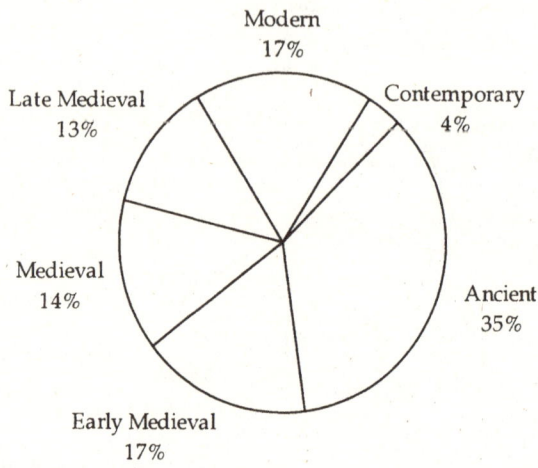

Figure 5.3: Distribution of Papers over Period Investigated

medieval, and colonial periods. But this shift in focus is not significant enough to indicate that a new generation of scholars with new
interests has entered the community. Out of the 404 papers published
between 1976 (10 years after the founding of the journal) and 1994,
the papers covering the ancient and medieval periods still account
for 51.4 per cent (208) of the total number, while the contributions
from the medieval, late medieval and modern periods total to 38.6
per cent. The corresponding figure for the period 1966–75 is 54.4 per
cent and 22.5 per cent. This clearly implies that there is an incremental displacement in interest from the ancient and early medieval
periods. At the same time, interest in the more recent period has
grown almost at the expense of the medieval period.

The predominantly researched themes can be inferred from the
two figures (Figures 5.2 and 5.3). Fifty-two (28 per cent) of the 200-
odd papers situated in the ancient period have to do with the history
of astronomy and mathematics in ancient India. If in addition to this
we cumulate the papers that address the history of astronomy and
mathematics from the early medieval, medieval and late medieval
periods, more than 50 per cent of the papers published are on the
history of astronomy and mathematics. Papers on the history of
astronomy and mathematics in ancient and early medieval India
account for 61 per cent of the papers published. This corresponds to
the epoch of the Vedanga Jyotisha, and the astronomers Varahmihira,

Aryabhatta, and Brahmagupta—iconic presences in the pantheon of Indian science. The smaller number of papers on the history of science of the modern period as well as the contemporary period betrays a lack of engagement of historians of science with the recent history of astronomy and mathematics, as well as the fact that the contemporary sense of the historical harks back over 1000 years.

At this juncture it is possible to infer something of the community of historians of science. First, the publications in the *IJHS* reveal the preponderance of historians of mathematics and astronomy. Further, these historians of astronomy and mathematics focused their attention on the study of mathematics in Indian antiquity. This either created the impression, or promoted it, that the history of science was more or less synonymous with antiquarian studies. And, finally, it appears that the historian of science in India, as elsewhere, bogged down in nationalist historiography, was lured by the idea of a 'Golden Age' of science and culture in India. Further, like its European counterpart, the history of science in India had to fabricate its 'Dark Ages', where science went into neglect. The explication of the onset of the Dark Ages provided a modality for deliberating upon the current state of science in India.

If we scrutinize the publications in the history of astronomy and mathematics, we notice a concentration of activity around the history of mathematics of the ancient and early medieval period. The focus on the modern and contemporary periods is negligible. The plot for astronomy is similar, except for the fact that there is a high peak in the region of the history of astronomy of the late medieval period. This is ascribable to interest in Zij astronomy, and the school of astronomy formed around the efforts of the 18th-century astronomer-king, Jai Singh.

However, an examination of the number of papers published annually indicates that on average, in a high-interest area like the history of mathematics, between two and four papers appear annually. Therefore, there are grounds for proposing that even the sub-discipline has not spread adequately in terms of either the number of new researchers coming in or in terms of researchers. In fact, sustained concentration around a set of chosen problems in the history of mathematics possibly indicates the operation of a close network of scholars. Consequently, the journal embodies an approach to the history of mathematics or the discipline at large. Following closely on the heels of the history of astronomy and mathematics is the history of medicine and the life sciences. In this case too the focus is on the history of ayurvedic medicine in the ancient and medieval periods, and touches upon unani medicine during the late medieval period.

Another feature worth exploring are the special issues of the journal. Over the last 28 years 10 special issues have appeared. These special issues contain papers presented at international conferences sponsored by INSA, that in turn funds the National Commission of the History of science. Table 5.1 lists these special issues.

TABLE 5.1
Special Issues of IJHS

Year	Special Issue
1969	The history of astronomy in india
1974	Copernican astronomy in its relation to other traditions of astronomy
1975	Al-Biruni in the history of science
1977	Aryabhatta in the history of mathematics and Astronomy
1981	The history of medieval medicine in India
1982	Science and technology in 18th- and 19th-century India
1985	The history of astronomy in India
1986	Ibn-Sina in the history of science
1992	Science in India, 1900–1980
1994	Calcutta: Science City

Five of these 10 special issues are on the history of astronomy and mathematics from antiquity and the medieval periods. From 1982 onwards, three of the five special issues deal with a recent segment of the history of science and technology. The 1992 issue was produced by the peers of science in India for each of the sub-disciplines. The purpose was a self-appraisal of each of the sub-disciplines of science in the contemporary period, as well as to historically situate the evolution of these disciplines. Since these accounts were produced by the scientific community, they would reflect the immediate concerns of the community and the posts occupied by various members of the community vis-à-vis their disciplines. Indeed, a notable feature is that the *IJHS* put technical education on the agenda of the history of science and technology studies in 1982. Historians of technique like A.K. Biswas had been chronicling the history of techniques in India in the journal.

A small number of articles appearing in the *IJHS* have raised historiographical questions and examined science from the vantage point of the philosophy of science. But most of these are efforts in translation underwritten by a canonized theory of science, that is either positivist or at best pre-Kuhnian. The aim of these exercises in interpretation is to tease out homologies between the epistemology of modern science and the knowledge systems, and practices of ancient and medieval India. From the perspective of the sociology of knowledge or cultural studies this programme might provoke a revolutionary critique; it must nevertheless be acknowledged that the new cultural discourses are themselves the product of a third world that is far more self-confident today. Further, in the West as well it is becoming increasingly difficult to sustain the mythology of the West as the acme of civilization.

Beyond the agenda drawn up by the founders of the *IJHS*, the papers themselves may be seen as responses to two ideological stimuli. In the early years, when science still reaped the cream of the idea that science was ratifier of a radical world order, there was an attempt to visualize science as the eventual victor in the battle between progressive ideas and obscurantist forces. However, Eurocentrism in the international domain, as well as the retreat of the socialist ideal in the 1980s, prompted a reverse commentary. This effort to translate the methodology of the knowledge systems of India into the methodological idiom of modern science and to reconfigure these knowledge systems in the light of modern scientific knowledge has been called the methodological imperative (Raina and Habib 1993). This imperative delimits the ongoing dialogue between modern science and other knowledge systems. The history of science, embedded either in sociological, philosophical, or cultural frameworks is yet to find an appropriate formulation and interpretive framework that would render this dialogue equitable and democratic.

THE RELATION BETWEEN SCIENCE AND ITS HISTORY

How is the object referred to as science visualized by those publishing in the *IJHS*? The answer to this question cannot be found through a tedious textual scrutiny of all the papers that have been published by the *IJHS*. For such an exercise would only reveal that no such generalization can accurately represent the picture. However, it is not too far-fetched to suggest in an environment where historians of science have emerged out of the community of scientists that there is a correspondence between scientists' conception of science and the growth of scientific knowledge and the perception of the very same object and processes within the community of scientists. This is a product of the institutional dependence of the history of science on the world of science. Hence, there are institutional obstacles that hinder the emergence of a critical theory of science and society, and historians of science would have to find ways of notionally liberating themselves of this institutional baggage.[6]

To decipher the nature of this relationship of dependency between science as practised in India and its historical reflection, we must take a look at science itself. In the post-independence era a charge commonly levied at the community of scientists is that their activity and programmes are drawn and defined by the centres of science in the developed world and hence is derivative. This peripherality of science in India engenders a crisis of legitimacy for science. It is this fracture in the social fabric that the official history of science seeks to cement. This role of the history of science cannot be ignored, since the National Commission for the History of Science is under the aegis of the Indian National Science Academy and does not come under the Ministries of Culture or Education or even Human Resource Development.

The extrinsic factors that shape the historiography of science in India are of a sociocultural nature and acquired prominence in the post-independence years. The histories that came to be rewritten during these years were at variance with the colonial and Orientalist narrative. Further, it involved an archaeological undertaking to 'bring forth' the buried traditions, knowledge, and nuggets of culture that had been lost to the scrutinizing gaze of the colonial historian. The history of science in India was thus part of the cultural project that emerged in the 1950s, wherein the historical narrative on science was but one element in a broader history of culture. Second, the project acquired cultural legitimacy in a decade when there was a proliferation of scientific institutions—for in the Nehruvian world-view science was the engine driving the development process.

The coordinates for the discipline were thus decided by the frame of scientism of the Nehruvian era and the programme of historical reconstruction that was sweeping post-colonial societies. Naturally, the neurosis afflicting the scientific community would find its reflection in the history of science writing. Furthermore, the global concerns of the scientific community would also strike a sympathetic chord among scientists in India as well as historians of science—thus, the history of traditional medical systems is presently a raging concern with historians of science. A feature worth noting is that while the theoretical concerns of STS discourse presently do not echo in the *IJHS* archive, the theory underlying the organization of this archive as well as its epistemology is structured by a positivist conception of scientific knowledge.

In an age where the social sciences have long abandoned positivism, and where even the Popperian conception of science stands challenged, it is a matter of wonder that such an orthodoxy continues to prevail in Indian scholarship, even though the conditions for its radicalization exist. The project, it may be humbly suggested, acquires legitimacy because it foregrounds the image of science and scientist, and presents this activity as a pilgrimage of truth. The project is guided by a commission that is supported by a scientific academy on whose governing council there has been little or token representation by a professional social scientist of repute (till recently)—and it is not that India lacks in such academics. Such commissions have been constituted of scientists with a passing interest in the history of science—a proviso that has not allowed for: (*a*) the professionalization of the discipline in 30 years, consequent to which (*b*) the field suffers from a lack of depth, save for the history of mathematics of the early to late medieval periods where there has been a concentration of effort. And even here it could be said that a sociology of mathematics in India has failed to emerge, a rather disturbing feature, given that India's accomplishment in mathematics is always mentioned with pride.

At the level of the theory of history, the scientistic account of science obviously validates the classical internal–external divide in the history of science. According to the tenets of this divide, there exists an internal logic of scientific development, a logic that is not tainted by external factors such as the sociocultural or economic. Accordingly, the external account suggests that the external provides a context for the kind of science that emerges, but here again the internal logic of the sciences is not determined by the social conditions that accompany its emergence. Society merely provides a backdrop within which science might fruitfully evolve or decline. Once science is accepted as a cultural universal, without any qualification, the need to critically examine the internal–external divide does not arise. The social serves as a backdrop for the staging of science, and is rendered active when the conditions for the emergence of the scientific revolution or its non-emergence are to be investigated.

Thus, the framework assumes: (*a*) that the internal–external divide is clearly defined; and (*b*) that the growth of science results in societal development as axiomatically given.

A framework that assumes a clear demarcation between the external and internal in science mobilizes the social to explain either the non-performance of science, or its inability to reach out to new audiences, or its inability to live up to its rhetorical promise to usher in the millennium. The highly limited horizon of the 'social' so constituted poses impediments in the path of the theoretical development of the history of science in India. The impediments are as much of a technical nature as they are of a disciplinary variety. This is partially because the history of science is considered a sort of amateur history: which implies that professionally the discipline lacks a set of agreed-upon techniques and a protocol for training. It also points out to the discernment of the nature of history within the scientific community: the history of science is seen as the production of a chronicle of scientific truth and reason based on the mobilization of textual or artifactual evidence.

As a result, the notable exceptions of Kosambi and Debiprasad Chattopadhyaya apart, the social history of science in India is underdeveloped. Debiprasad Chattopadhyaya's effort was to identify the obstacles in the path of the emergence of a full-fledged scientific tradition in India. This history was transcribed as a battle of ideas, with two contending camps, that of materialism and its confederate science, and idealism and its ideological confederate religion. The history of the losing battle fought by the materialists in India stands as an epitaph to the non-emergence of modern science in India. Chattopadhyaya's efforts are much better appreciated through his sizable collection of books rather than through the modest publications in the *IJHS*.

The conventional philosophy of science that was founded on the Whewellian idea that science was endowed with an epistemic engine capable of generating the laws of nature was assumed as an unproblematic given in the *IJHS* archive.[7] As the philosopher of science Stove once put it, the scientific community the world over is Popperian in its appreciation of the growth of scientific knowledge. But the manner in which the richness of Popper's ideas contributed to their own demise is now a matter of history (Allport 1991). This revolution in science studies amongst scholars studying the sciences and their histories is not evident in the papers published in the *IJHS*. The object of investigation called science appears to be frozen in time and eternity. Paradoxically, the historians of science publishing in *IJHS* have been ahistorical in their appreciation of the representations of science.

The historiography that blends different perspectives into an *IJHS* archive is based on a set of disciplinary and cultural valuations. One of its research programmes or correlates in the history of science finds expression in transmission studies. The fundamental postulate that motivates transmission studies claims that scientific knowledge is transmitted from regions of high truth to regions of low truth concentration (Shapin 1983), and this knowledge system and its nested value system when transmitted is unattenuated by the novelty of the cultural system into which it is implanted. The primary objective for transmission studies is the identification of the apostles of science and a narrative of conversion that draws the ignorant into the cosmology of science. The cultural milieu is considered passive and so are the converted. This version of the history of science, that shares much in common with imperial history, is common to historians and historians of science who write the history of science from a Eurocentric vantage point, as well as for historians who are uncritically convinced that science implicitly embodies what Lyotard (1984) calls a 'narrative of freedom', irrespective of the spaces and historical epochs that determine its specific incarnation.

Nevertheless, it is essential to differentiate between the two senses in which transmission appears in the history of science studies. In the first case it denotes cultural transactions across civilizations in antiquity and during the medieval period. Since scientific knowledge and rationality from the age of pre-modernity are not privileged either culturally or politically, 'transmission' lacks the evangelical connotation that it conveys in the history of science studies of ages where it is either coupled with nationalism or provides legitimacy to the doctrine of progress.

In the second sense transmission studies provide a frame of research for the study of science at the periphery of the metropolises of science. Methodologically, this involves assigning priority to a knowledge claim, especially those central to the growth of modern science and technology, and then identifying the channels of communication and percolation of this knowledge. Implicitly, there appears to be a tradition of high science and a tradition of low science. And the agenda of transmission is to weave the globe into an international community sharing common norms. Research within this genre of historiography problematizes neither science nor a theory of communication, and comes up with a series of evaluations, some of them very deeply nested.

I have argued elsewhere that transmission studies, studies on colonial science (Bassala 1967), and certain variants of the centre–periphery model share a common ground and can be said to constitute a family of overlapping historiographies.[8] Nested within these are the assumptions of nationalist historiography. These constitute the foundation of the *IJHS* archive. It has been pointed out by historians of science elsewhere that the history of science cognitively informs science education (see the papers in Shortland and Warwick 1989). Further, current scholarship in cognitive science has established that the trajectory characterizing the evolution of a scientific theory bears a remarkable homology with the learning curve of students—both are earmarked by similar learning stages. The history of science thus directly interlocks with the programme of science education and teaching—the recognition of which comes with the founding of a journal like *Science & Education*.

Societies that have had a colonial past must confront the dilemma of legitimating scientific knowledge. Modern science discussed in the context of its origination is often presented as a break with the knowledge systems or the cosmos of medieval Europe, and yet when considered in its relation to non-Europe, is re-presented as a direct progeny of a miracle that occurred in ancient Greece. Colonial and post-colonial societies trapped between conserving their knowledge systems and adapting to modern science have frequently had to address the cultural practices of the new knowledge form head-on. This means that efforts to institutionalize and legitimate the new knowledge form would have initiated processes of cultural appropriation.

In historical consciousness these responses are manifest in interrogations that look upon the past in revivalist or revitalist terms. Nationalist historiography in order to create this legitimate space for science evokes the past of India as a 'me-too' or 'we did it first'. This unfortunate aspect of colonialism and the imperative desire to neutralize a pervasive Eurocentrism has restricted the likelihood of the emergence of an ecumenical version of the history of science in India. Such an eventuality tends to shift the focus of investigation away from the study of social processes underlying the generation of knowledge to a fixation with 'priority disputes' or examining the social conditions responsible for the non-emergence of modern science in India.

The impression that the history of science in India is yet to find an anthropological or sociological voice hardly needs elaborate demonstration as far as the *IJHS* is concerned. But more importantly, this circumstance itself is evidence that scientific and technical knowledge is conceived as inhabiting a textual world, further that these are articulations of high culture. In addition, the fixation for the mixed sciences or a textual reading of science has also produced a gross neglect of the history of techniques and technology. Extending beyond the *IJHS* community, it may be suggested that the history of techniques and technology is a woefully underdeveloped field of study in Indian today. As long as the peers of Indian science and technology are themselves partisan to the ideological representation of technology as applied science or a lower kind of knowledge, this state of affairs is bound to continue (for an epistemic critique of this fallacy see Vincenti 1990).

The uncontested character of science itself as an object of investigation and the crucial obliviousness to the social character of the production of scientific knowledge or techniques has induced a theoretical dependency on two fronts. On the one hand historians of science and technology have an eidetic commitment to the reconstruction of the history of science that is framed by the Eurocentric account of science, although they may disidentify themselves from it. The narratives generated within this genre are either counter-narratives or at best appendages or subtexts of a Western account of technology. Second, a product of the placid acceptance of science as a neutral given is the revivalist reading of the history of science. The emergence of Hindu revivalism in the political space finds an echo in the revivalist edge to some of the papers appearing in the *IJHS*. Epistemologically and politically they exhibit a medieval closure of the horizons of science and politics, a closure that the epigones of Galilean science had sought to eliminate. This trend is also emblematic of the new political formations seeking scientific legitimacy, and the subsequent mobilizing of the scientific community towards that end.

PHILOSOPHICAL APPROACHES:
THE SEARCH FOR COGNITIVE HOMOLOGIES

For those predisposed to anecdotal history and looking for origin myths, Nehru's incarceration in Naini Jail in the 1930s may be considered for some the beginning. His reading list during the three years spent in jail includes all the canonical books authored by the leading members of the Cambridge left as well as Fabian socialists. However, from the perspective of science and politics, we begin to see a steady alignment of two trajectories—of a series of deliberations on science and planning emerging in the 1920s within the science and culture group in India, and Nehru's gradual mobilization of a network of scientists working in India in league with some of their colleagues at Cambridge. The subsequent trajectory of the institutionalization of science during the late 1940s and 1950s was more or less framed in those years. And it is interesting to see how the image of science that coruscates in the accounts of historians of science during the 1950s and 1960s closely follows on the heels of the science and planning discourse of those years.

In epistemological terms, some of the articles appearing here aim at teasing out homologies between the epistemology of modern science and the knowledge systems, and practices of ancient and medieval India.[9] One of the legacies of the Orientalist tradition was the predisposition of the *IJHS* historians to direct their gaze to the Indian philosophical systems in order to divulge the origins of science in the Indian tradition. The internalist tradition, guided by a philosophical conception of science, thus was led onto the question of the foundations of science in ancient India. Six principal philosophical schools defined their discourse: Purva Mimamsa, Uttara Mimamsa, Samkhya, Yoga, Nyaya, and Vaisesika. The historian of science Subbarayappa (1967: 2) interpreted the Vaisesika as a pluralist realist system, considered as the school of Indian atomism.[10] Let us examine this construction a little more closely. The question that needs to be asked is how historians of science recover the Vaisesika. Since the British Indologist A.B. Keith's *Indian Atomism* to contemporary readings of the history of sciences in India, cognitive homologies between several strains of Greek atomism and Vaisesika atomism have been proposed by a number of historians of science. Historians of science from the mid-19th to the middle of the 20th century had celebrated the rapid evolution of modern atomic theory as a theory of physics. This celebration must be situated against the backdrop of the polemic between field theoretic descriptions of physics and the corpuscular description. In much of this historical narrative the evolution of atomism is presented as the victory of materialist thought. From the perspective of origins, a lineage, however complex, was drawn from Greek philosophical atomism to modern-day atomic theory as a sub-discipline of physics.

Now, if the Vaisesika system was ontologically committed to atomistic realism, it became important to ask why it failed to blossom into a theory of the physical sciences or to stimulate the growth and elaboration of the sciences in India.[11] Subbarayappa proposes that there was a conceptual transformation of the school from a heterodox one into an orthodoxy, and this had to do with the synthesis of Vedic ideas that were integrated into the Vaisesika opposite. And though the Vaisesika was subsumed within the orthodoxy by the 9th century, its epistemology required that perception and inference were important instruments for the acquisition of knowledge, which in turn furthered ideas both 'naturalistic and rational' (Subbarayappa 1967: 23). An over-determinationist framework prompted numerous responses to this core interrogation.[12] If Subbarayappa and other contributors to the *IJHS* history were seeking answers in terms of the cognitive or epistemological transformations of philosophical systems themselves, the Marxists were attempting to reveal the sociopolitical basis of orthodox thought, and the legend of suppression (Raina and Habib 1999) provided them with an appropriate explanatory device. It is important to note that while over-determinationist history restricted both the kinds of questions that could be asked and the responses proposed, it did further important insights through the perspective of cross-cultural history. With the benefit of hindsight, it may be remarked that the cross-cultural studies of science reckoned with the limitations of over-determinationist history.

Returning to the reconstruction of the Vaisesika, we can observe the appearance of diachronic maps or tables listing conceptual landmarks in the history of four major civilizations across time. These maps or tables served two related purposes—establishing priority, who did what first, and once this had been demonstrated, charting out possible channels for the transmission of ideas. Thus, for Subbarayappa the significance of the Vaisesika resides in the fact that its theories of elements and atomism were integral parts of a larger philosophical system. This was another way of proposing that the system had developed autochthonously or internally. Further, the school is supposed to have come on to its own well before pre-Socratic ideas had taken on definite shape (Subbarayappa 1967: 30): In which case, the theory of atomism could not have been developed as a consequence of transmission from Greek sources. These histories from the 1950s and 1960s argued from a singular epistemic conception of science and thus Subbarayappa (ibid.) writes: 'If the Greek view is rational so is that of the Vaisesika for there is a theory of causation running through the latter.' In non-Western nations then, particularly those that have had a colonial past, there is a perceptible subterranean discourse that seeks to reverse the amnesia that characterized the history of science prior to the 1950s—however, this reversal was framed in the very same mould of the history of science that characterized the former, namely, the empirical–mechanist scientism of the era.

This genre of the comparative history of science was structured by two propositions in turn founded on the ideal of science as a cultural universal: (*a*) that the so-called scientific method had remained more or less constant over the centuries; and (*b*) that the structures of pre-modern knowledge systems were unproblematically apprehensible in the light of that of modern science. The latter proposition in effect mediated the comparisons between the knowledge systems of distinct culture areas across time and space. In other words, constitutive of modern science was an objective and neutral translation machine or epistemic engine. As a result, there was no incest taboo that prevented the inverted reproduction of the chauvinism underlying the Eurocentric history of science. Subbarayappa has never been open to the charge of Indocentrism; but the bug of priority, so foundational to the nationalist historiography of science as much in the West as in the East, rendered it difficult, if not impossible, to decipher when it was cognitive justice that was being sought after, and when the project was lapsing into a new centrism.

Let me spend a little more time on this philosophical reconstruction of science founded on the script of science as a cultural universal. Take the case of a philosopher and not a historian of science with a background in the sciences (most often physics in India as elsewhere, I guess)—I consider a well-known philosopher but little known historian of science, T.M.P. Mahadevan. Mahadevan shares a programmatic stance with the founders of the discipline of the history of science in India such as B.N. Seal, and which runs through the discourse on science and culture during the first decades of the 20th century. During this period the advocates of science adopted a diverse range of strategies in order to ensure the successful grounding of modern science within Indian society—a colleague and I have elsewhere have called this the programme of critical assimilation (Raina and Habib 1996)—which in turn spawned a number of historical investigations on the sciences of India.

These reconstructions bear a close affinity with those of the logical positivists. The most conclusive of these reconstructions assert that of the knowledge forms of India, the disciplines of medicine and linguistics were the only ones that shared a methodological affinity with that of modern science. Furthermore, irrespective of the contributions made to science in areas such as astronomy and mathematics, in the present context it was Indian philosophy that could contribute to modern science. This, it could be conjectured, was a spin-off of the quantum revolution in physics, wherein physicists such as Schrödinger and Pauli had attempted to guide their deliberations on the nature of quantum reality based on their manner of apprehending Vedantic and Buddhist philosophy. Thus, Mahadevan, much in the fashion of the Vienna Circle, and not very differently from what cognitive science does today, argued that Indian philosophy could intervene in the production of modern scientific knowledge, particularly within the domain of logic and methodology, the analysis of language and meaning, in elaborating upon the relationship between branches of knowledge and their unification, and in highlighting the place of experience in the generation of knowledge (Mahadevan 1969: 27). The last was introduced to disidentify with the Western ascription that Indian philosophy was other-worldly.

A feature of this positivist reconstruction, as different from the Marxist scientism, for which even in antiquity there was a one-way arrow leading from atheism to materialism to science, was that it never shied away from the circumstance that in antiquity scientific notions could have emerged in a sacred cosmos. Mahadevan's metaphysical pragmatism bears much in common with 17th- and 18th-century naturphilosophie—the investigation of nature, as an Indian sage Aurobindo put it, was a preparatory stage to the learning of the higher spiritual truths (ibid.: 28; Raina and Habib 1996). Similarly, the need to improve the precision in measurements of distance and time was driven by the doctrinal need to secure the efficacy of ritual incantation. This was pointed out by Thibaut in his work on the Sulbasutras. Within the same context, there was much concern whether the Indians knew about the Pythagorean theorem. Western scholarship had deified the notion of 'geometric proof' amongst the Greek mathematicians, and then projected it as one of the central elements of Greek science that was later to become a constitutive element of modern science. Nineteenth-century history of mathematics had branded the mathematics produced in India and China as algorithmic or algebraic. In order to neutralize the judgmental intent of this reading historians of science in India and other parts of the world sought to show that equivalents of the Pythagorean theorems were to be found elsewhere. However, what has been settled with certainty is that the property of Pythagorean triples were known amongst several ancient peoples long before Pythagoras proved this theorem on Euclidean lines.[13]

In arguing the thesis that the sciences of antiquity were situated in a sacred cosmos or a *scientiae sacra*, even the study of language and meaning or the science of linguistics in antiquity served as a *sadhana*— or a penance—in order to ensure the realization of Brahman (Mahadevan 1969: 29).[14] Further, in the case of medical practice in antiquity, it was educed that the discipline could not decouple the self (the mind would be a Cartesian category) from the body. The self was primary, and this self was situated and united in a mind–body complex (*samyoga-purusa*) (ibid.: 30). The efficacy of a cure or a curative regimen was dictated as well by the practice of an ethics that bound both patient and doctor (ibid.). In this writing of the 1960s and early 1970s, we encounter as subterranean texts anxiety concerning the moral vacuity of science, the pandemic condition of industrial civilizations—and these anxieties figured in different ways in the constructions of the imaginary past.

As mentioned earlier, even in the more formal logical disquisition there was an attempt to retranslate the methodological precepts of Indian philosophical systems on lines homologous to that of modern science (ibid.: 31).[15] Such interpretative exercises were obviously founded on the postulate of the methodological unity of the sciences across time—and this is possibly the most central presupposition of the philosophical approaches. Slightly beyond the ken of the *IJHS* there lingered the parochial reading that all of modern science prefigured in Indian antiquity. In the 1970s, when the arch of scientific knowledge began to crumble, in India revivalism intersected with relativism to assert that each culture had its own science endowed with the ethos of that culture, and that there was little scope for a dialogue across cultures.

The philosophical tradition ran its course throughout the 1960s and 1970s, quite oblivious of the Kuhnian revolution that had produced a reconceptualization of science and its dynamics. The philosophical approaches were enabled by a tradition initiated by the five volumes of *A History of Indian Philosophy* authored by Surendranath Dasgupta. Dasgupta himself, while trained 'traditionally', sought to translate the Indian philosophical tradition in terms of the concerns of Western philosophy. He adopted as his exemplar the German scholar Windelband's *A History of Philosophy*.[16]

Radhakrishnan's two-volume *Indian Philosophy* could be seen to belong to this very tradition. However, by the 1960s, 150 years of Oriental scholarship, as well as the efforts of Indians embarking on these projects of translation were to transform these philosophical texts from the past in a manner in which it was no longer possible to recover their original sense, if there was one, again. In any case, the primary purpose of these philosophical excursions into the history of science were three-fold: Indian philosophy was to provide a meta-logic for the sciences—an idea possibly inspired by the Vienna circle; this would in turn facilitate the redrawing of disciplinary boundaries; and, finally, philosophical excursions into the history of sciences would contribute to the development of an ethics for a science that may have lost its soul (Mahadevan 1969: 40).

A LOST OPPORTUNITY FOR THE HISTORY OF SCIENCE

Shifting our focus within the same tradition, we turn to the other end of the temporal and conceptual spectrum. This relates to the history of the modern sciences in India, inaugurated during the period of the East India Company, and subsequently with direct British Rule. S.N. Sen, one of the co-authors with Subbarayappa of *A Concise History of Science in India*, was actively involved, albeit unsuccessfully, in giving the discipline an institutional identity. Sen began by studying the introduction of modern sciences in India that commenced in the late 16th century with a trickle of medical men from Europe, naturalists, Jesuit missionaries, and adventurers. What began as a trickle became a steady flow of scientists, doctors, and military engineers. Sen (1988: 112) went onto pose the question that I should like to call Sen's question: why, despite close and long contacts with European science, was the introduction of modern science in India so slow till the closing years of the 19th and early decades of the 20th century? The question becomes significant because this was the century characterized by the most expeditious expansion of European colonialism as well as by the diffusion of European science.

Sen was intrigued by the fact that between the trickling in of modern scientific knowledge and the first publications of Indians on the so-called modern sciences almost a 100 years were to lapse. And so he proceeds to examine: (*a*) *internally*: the character of science transplanted in India, and the motivations behind this programme; and (*b*) *externally*: the social and psychological preparation necessary for the assimilation of the new knowledge. Such an investigation necessitated sensitivity to the difference in emphasis accorded by Europeans and Indians to scientific progress (ibid.: 113). On this count Sen mustered adequate historical evidence to establish the point that European efforts during this period were restricted to 'the field sciences': geography, geodesy, geomagnetism, geology, botany, agriculture, and astronomy. The principal agenda was to chart the resources of the Indian subcontinent in the hope that the appropriation of resources be rendered more efficient. The circulation of this knowledge was also tardy on account of the fact that these sciences and their findings were shrouded in secrecy, since scientific information so obtained was essential to the competition between the colonizing nations of the time (ibid.: 118).

The activity of the Jesuits in India, as Sen (ibid.: 114) was to observe, 'was particularly barren and abortive: there is nothing to show that the Jesuit contacts facilitated in any way the transmission of European science to India'. Within the perspective of comparative history, the history of the transmission of European science in India was entirely at variance with the Chinese one. Sen's reading was premised upon his conception of science, and his conception of what was a scientific contribution. On the one hand, like most historians of science of his generation, physics was the exemplar of the sciences. And surely the Jesuits in India did little physics, though that was not true of astronomy in the 18th century. But what the Jesuits did do was undertake a monumental task of preparing dictionaries of some of the regional languages of India, such as Kannada, Telugu, and Tamil; while Sen was cognizant of this work, he doesn't address the importance of this enormous compendium—possibly leaving it to the historians of 'culture'. Second, in the later half of the 19th century the Jesuits contributed no papers to scientific journals, but their contribution to the cause of education in general and science education in India is undeniable.

The British delayed the introduction of research in to the university charter for almost 50 years. Sen and subsequent generations of historians of science in India examined the development of the scientific research system, particularly in domains distant from the field sciences in the concerted efforts of scientists trained in India and abroad. These scientists, driven on by a nationalist agenda pressurized the imperial administration into including postgraduate teaching and research in the agenda of university education (Sen 1988: 119). Sen then turned his investigative focus to the study of the history of education in India, commencing from antiquity down to the period of British rule, in order to acquire an understanding of how the legacy of educational institutions was transformed, erased, and substituted with the onset of colonial rule in the hope that this would possibly reveal the process of assimilation. As long as the history of science was still in its scientistic mode, it was difficult to study assimilation in terms of the dialectic of the traditional and the modern. The historiography of transmission could best be exemplified in the metaphor of the percolation models. We will come back to that later. But the core hypothesis around which Sen's research was organized was that the British educational experiment was grafted upon a culture possessing educational institutions for roughly 2500 years (ibid.: 1).

Sen chronicles how most institutions for the dissemination of scientific knowledge were associated with temples of various denominational orders. This reading was founded on the assumption that all such educational institutions on the Indian subcontinent were organized and managed along similar lines as those of centres of learning located around the temple cities of medieval south India. In fact, these institutions were thus seen as: (*a*) continuing in a tradition from hoary antiquity, that in turn (*b*) were uniformly distributed throughout the country, and (*c*) their fossil remnants were still to be found frozen at the time of the onset of colonial rule. Two historiographic considerations more or less impelled this reconstruction. The first had to do with the greater quantum of physically inscribed historical evidence—be it artifactual or in the form of monuments in southern India as opposed to the north (Sen 1988: 8). Second, in the just turned sovereign republic India, national identity could be constructed around the idea of the unity of cultural institutions since the nation was a jigsaw of linguistic and ethnic entities, and educational institutions were just one of the many.[17] Furthermore, his reconstruction of these institutions of higher learning bears many affinities with that other medieval institution now distributed all over the world, namely, the medieval European university, probably in turn inspired by the idea of the Baghdadian Bait-ul-Hikmah. Sen elucidates that the focus of pedagogy at these temple-associated state-supported institutions qua universities was on formal knowledge, rhetoric, and aesthetics. And as has happened so often when writing Indian history, the reconstruction is often based on foreign sources. This time Sen (ibid.: 9) indicated that the textual evidence for the reconstruction was entirely from Tibetan and Chinese sources.

As distinct from the formal disciplines, technical education and the system for imparting training in the arts and crafts came with contractual obligations binding on both master and the apprentice. That Sen himself was situated on the trajectory of the history of science and technology of his times is reflected in his conclusion that for all the virtues of the system, "twas beset with drawbacks inasmuch as it was not conducive for the industrial revolution' (ibid.: 19). The idea of innovation, so important in subsequent history of technology studies, figures there albeit in a preliminary form.

It is evident that across the historiographical divide, internalist–externalist, Marxist–non-Marxist, the traditional systems of medicine in India were the closest to approximate to the tenets of modern science (Sen 1988: 21). Thus the institutional and philosophical accounts were in concurrence. It is important to understand why this was so. Most medical texts from ancient India were prefaced by a methodological chapter where the central role of experience, experiment, and logical reasoning was emphasized. In more ways than one, the schools of logical realism such as the Nyaya–Vaisesika informed these traditions particularly in the study of etiology of disease—for example, how was the accompanying or concomitant cause to be separated from the instrumental cause (Chattopadhyaya 1979). This clearly defined regimen articulated in the language of experience and reason offered itself as the prototype of a scientific theory.

Further, Sen discusses the educational institutions that appeared during the medieval period, these being the *maktabs, madrasas* and the *karkhanas*—in the case of the former two, in addition to a religious education, a scientific education was also imparted. The Arab synthesis of several scientific traditions, and its equally important role in the emergence of scientific revolution, led historians of science studying the medieval period, Sen and Rahman included, to place special emphasis on scientific learning during the medieval period (Sen 1988: 23). However, this overlooked the place of science in medieval culture. First, during the medieval period science was not at the centre of culture, a place it did move to in the 20th century. Second, science in the contemporary context is in a very different relationship with other branches of learning that must be explicitly recognized. Hence, any past recuperation of science that foregrounds it overlooks the complex encystment of different knowledge domains.

One aspect of Sen's studies into educational institutions by far outweighs those relating to, say, the history of astronomy or that of the introduction of modern science. This has to do with the proliferation of educational institutions throughout the Indian subcontinent from the medieval ages. This runs counter to the picture created by British historians that provided one of the legitimatory tropes of British rule: that it was the British Imperial Government that lit the lamp of enlightening knowledge on a dark continent plagued by superstition and ignorance. Officials of the East India Company (Adams 1865) noticed the broad expanse of educational institutions just before the onset of British rule. The importance of this historical feature in rejecting many of the claims of imperial history that were foundational for the Eurocentric history of science was not picked up by Sen's generation of historians. However, historians of science working on Egypt and the Arab world today are coming up with very similar conclusions (Crozet 1999) in their respective contexts. And this has meant that educational practices in turn have to be situated culturally. This takes us back to the dialectic of tradition and modernity. The new dialectic would have to disavow itself of the notion of tradition, being both a burden and obstacle to the development of knowledge. But then that would be a different historiography and programme. But it still remained for Sen to explain why this educational system either failed to evolve or was replaced. Situated favourably on the horizon of modern science, Sen (1988: 26) writes that for some decades prior to the institution of British rule, traditional institutions had fallen into decline, and this decline followed in the wake of the disintegration of the Mughal empire. In a paper he wrote in the mid-1960s, Sen (1966) recognized that the history of sciences was undergoing a tremendous change, and this change was of the nature wherein the social determinants of science were going to become more important. But the brand of externalism Sen practised during these years was none other than internalism. I would not be stretching a point too far in suggesting that in an environment where historians were nested within the community of scientists, there was a neat correspondence between the scientists' conception of science and the growth of scientific knowledge, and the perception of the very same object and processes by the community of historians of science. This was a product of the institutional dependence of the history of science on the world of science. Hence, certain institutional and cognitive barriers hindered the development of a critical theory of science and society.

THE END OF THE GOLDEN AGE OF SCIENTISM

While the *IJHS* genre of the history of science owed much to the Orientalist archive in terms of recreating the high traditions of Indian civilization, they were at variance with the colonial one. In fact the *IJHS* project was, metaphorically speaking, an archaeological undertaking to bring forth the buried traditions, knowledge, and nuggets of culture that had been ignored by the scrutinising gaze of the colonial historian. The *IJHS* history was thereby part of a cultural project that emerged in the 1950s. The project itself acquired a station during that period of the Nehruvian era when there was a proliferation of scientific institutions.

The scientistic frame was adopted for the history of science studies by the *IJHS*. One must also reckon with the fact that this was the era when the two-cultures debate drove a deep wedge into academia: wherein the culture of science masqueraded as a 'priestly culture' (Wertheim 1997), and the scientist was seen as the generator of truth. The history of science project was led by a commission supported by a scientific academy on whose governing council there was little or token representation by a social scientist of repute, or for that matter any social scientist at all.

There were serious research attempts undertaken to respond to the Needham question. But given the accepted terms of the internal–external divide, society provided a backdrop within which science might fruitfully evolve or decline. If science was a cultural universal without qualification, the social was merely an arena for the staging of science. The social was rendered active when the need arose to explain the non-performance of science or its inability to reach out to new audiences, or finally when science failed to live up to its rhetorical promise to usher in the millennium. The highly limited horizon of the social so constituted posed impediments for the theoretical development of the history of science within the framework of the *IJHS*.

This view of history was over-determinationist, as the scientist's history of science tends to be. This brings us to the paradox that while these historians professed inspiration from Needham, they remained within the over-determinationist frame. As Fuller (1999) points out in a recent paper, Needham challenged the existing historiography of his time by insisting that history mattered not for the sake of the past or for the sake of history, but for science itself. This possibility could not have been appreciated for those situated on the frame of triumphalist science—though one may well ask whether the locus of these Indian historians as former colonial subjects may not have sensitized them to this possibility.

However, we cannot be unduly critical of the historians publishing in *IJHS*, for in a way they pioneered a discipline and made difficult choices in an environment where to leave science to take up its historical study was considered the option of failed scientists. Further, even within their internalist framework, they did indicate how treaties on Indian philosophy and linguistics could inform modern science—but even that was an idea that could have achieved realization had somebody from the scientific community decided to take it up.

A narrow shift in focus from the history of science to the history of education momentarily opened the window that might have altered the picture of the history of science by questioning its foundations. However, so prevalent was the evangelical ideology of science that the importance of some of these findings were neither picked up by the historians themselves nor by their audience. Only when science came to be decentred from its privileged place in culture were the implications of some of these findings picked up within the framework of the critiques of modernization. The school remained fairly restricted, and the Marxist historians of the time such as Kosambi and Debiprasad Chattopadhyaya toiled outside this framework.

INSTITUTIONAL BARRIERS AND
THE FUTURE DEMANDS ON THE DISCIPLINE

A number of institutional obstacles, some of which are of a purely conceptual nature, have arrested the evolution of the discipline of the history of science and technology in India. When we speak of institutional we do not merely refer to academic institutions where different academic or research interests are pursued, but equally to the recognition of disciplinary boundaries, funding, and review practices. Thus, while science as practised might be distinguished by a great deal of permeability across disciplinary boundaries, within the institutional sphere this permeability is absent.

The discipline of the history of science is one of the casualties of a double bind. For the community of scientists the discipline is not 'hard' to merit support from the technocracies supporting the 'hard sciences', save as a public relations enterprise or to pander to the preoccupations of scientists with their place in history. On the other front it is not 'soft' enough to be embraced by the social sciences or human sciences for two possible reasons. The object of inquiry is the history of the hard sciences, in which case it should be an appendage to the hard sciences. Second, during the formative years of the discipline in India, it was dominated by internalists or those historiographically committed to empiricism, and their concerns never interlocked in any tangible way with that of the historians. As a result, it may be conjectured, the discipline was itself institutionalized at a time when the two-cultures ideology was still triumphant. As a result science was projected as a universal that sought to finally transform even culture along scientific lines—this imperialism ensured that the history of science was distanced from the social sciences.

Assuming that there was a relationship of supply and demand between the sciences and the history of sciences as academically ensconced disciplines, then the latter, if supported from within the sciences, must pay the price of leading a subordinate existence. In addition to providing a self-congratulatory account of the growth of scientific knowledge, the history of science can and must also deliberate upon the theory of knowledge produced by the community. In the last few decades it appears that mission-oriented science has grown at a faster rate than science in the academic sphere. The history of science as a discipline that entertains a theory of the production of knowledge within the portals of a university or an 'academy' must under the present circumstances confront novel institutional arrangements which in turn requires a revision of its own theoretical presuppositions and questions. The concomitant metamorphosis of the image of science as knowledge has not become the subject of analysis.

The hypothesis that mission-oriented science has grown at a faster rate than science in the academic sphere is evident from the marginalization of the university as the site for the production of knowledge. And yet it must be stated that the budgetary allocation for scientific R&D in the country is measly—under 1 per cent of the GNP. But the asymmetric growth is reflected in the perception that science is a sort of elite activity—a perception that the community has done little to dispel. Ideologically, then science is implicated in an orthodox relationship with the scientific community, in contradistinction to the social sciences where research pedagogy promotes critical inquiry.

Further, unlike in the age of early modernity, science is no longer in an antagonistic relationship with the state. The modern state to legitimate its existence as 'modern' must champion the cause of science, and through a fairly obvious process scientists and politicians enter into a relationship of exchange (Weingart 1993). Under such circumstances a critical appreciation imperils the community's agenda, while a linear, simplistic, adulatory narrative is the only one that could promote its interests. I do not think that this is true only of India. The history of the history of science in Europe and the West appears to have passed beyond this phase; but that it did flounder in the early history of the discipline is cogently expressed in the following words of Ravetz (1992: 148):

The ideological motivations and functions of various scholarly disciplines that study the natural sciences have had a shadowy history. On the one hand the glorification and defense of science in general, or loyal praise for the founding fathers of some particular specialty, were quite legitimate concerns of the (usually amateur) philosophers and historians of science until recently. But since they nearly all shared in the defining assumption of our science, that it is simple truth, having no connection with ideology, they could not rise to self-awareness about their own efforts. Only now, given the combination of professional self-consciousness among some scholars, with the current mood of disenchantment with science, can a truly critical analysis of the past and of the present be achieved. This means that the way is now open for a genuine history and sociology of science: for until such reflective disciplines have some critical distance from their object, they cannot produce anything more than anecdote, chronicle or hagiography.

In the present context where science needs little promotion of the medieval variety, the historian of science has a moral responsibility to the public and society. On the one hand there is a need for a critical appreciation of science and technology, for as Collins and Pinch (1993) have lucidly illustrated, this science and technology is not a disembodied object, but a social one, that in fact it is 'our science and technology'. The historian of science must then reach out to the scientific community and the public and establish a dialogue between the two. This would enable the scientific community to re-examine their own activity in a manner that specialization has thus far prohibited them from doing. This might also mean combatting the anti-science movement at the level of social movements.[18] But even here an equanimity in perspective might be warranted.

Looked at another way, the anti-science movement may be construed to be society's way of culturally appropriating the new knowledge (Elzinga and Jamison 1981), and on this count both the scientist and the historian of science would have to acquire new sensibilities and sensitivities. What then is the purpose of history concerned with process and of going beyond 'anecdote, chronicle or hagiography'? The history of science must contribute in the final analysis to a theory of science and society, its role in history, its relationship to culture, and values. These theories would have a bearing upon science and social policy. A more modest aspiration in an era where science is breeding new formalisms is to capitalize on the potential for redrawing disciplinary boundaries. STS might offer the broader framework for doing so, the history and philosophy of science and technology may provide the conceptual framework for the enterprise.

NOTES

1. Two core propositions of a methodological nature that defines STS as pursued in university departments are: (*a*) the denial of only an internal history of science; and (*b*) ethnographic practice enjoys epistemic privilege in the field (Fuller 1993: 7).
2. See Chapter 2 in this volume.
3. Roshdi Rashed (1994) remarks that Western scholarship on the Arab sciences exhibits the presence of Greek influences and has disregarded original Arab texts on the sciences. This would not be entirely true of the Indologists, for some of them did a lot more than translating texts reflecting Greek influences. It was Colebrooke who brought Bhaskaracharya and Brahmagupta to the notice of 19th-century mathematicians and historians of mathematics. Thus, while falling in line with the broad argument of Said's Orientalism, what would be required is a geographic typology of Orientalist texts that is flagged by the political posts occupied by the producers of these texts.
4. See Chapter 3 in this volume.
5. A study of the theories of the history of science suggests that with the cleavage between science and philosophy in the 19th century, internalism emerged as the institutional embodiment of the ideology that 'if science gave us the most comprehensive grasp of nature, then the most comprehensive grasp of science could be gotten by studying science scientifically' (Fuller 1997: 7).
6. In an important paper, Shapin (1992) has highlighted the nature of the assymetrical relationship between the internal and external history of science, and that one of the legacies of this dichotomy is that the latter is condemned to ride piggyback on the former. For a discussion of the reassertion of the internalist self-image of science during the Cold War, see Fuller (1992).
7. The most sophisticated formulation of this notion, and its logical culmination, is to be found in the oeuvre of Popper.

8. See Chapter 7 in this volume.
9. I use the term homology in the sense employed by the Japanese scholar Kaneko (1987: 359), although it is a usage that is circular. Cognitive homologies refer to cases where 'an isomorphic state of mind or similar mentaility can also be found in the semantic sphere of another culture'.
10. Most redactions of the Indian philosophical systems interpret the Vaisesika as a school of realist atomism.
11. The question as posed by Subbarayappa (1967: 31) was: 'Why was it that the Vaisesika school with its presumed heterodox beginnings and owing allegiance not only to the reality of the external world but also the relationship between cause and effect was not able to gather momentum and establish a scientific tradition as we understand it now?' As is evident, the question is entangled in a maze of historiographical wrangles, the most amazing feature of which is the triumphalist vision of modern science, more or less dictating the question posed by the historian.
12. According to the over-determinationist perspective of history, 'there is only one world order, which consists of certain nodal events through which all possible histories would have had to have passed' (Fuller 1993: 210–11). On the other hand in the under-determinationist view 'a given event need not have occurred, but once it did occur, everything that followed did so by necessity'.
13. Two observations could be made here. First, a historian of Chinese mathematics, Karine Chemla, to break out of this Grecian notion of proof, has proposed in her recent papers a conception of proof to mean any mathematical procedure employed to obtain a solution. This draws further support from the new mathematical procedures that have emerged at the interface of computer science and mathematics. Second, with the upswing of number theory and algebraic geometry, whether Indian historians of mathematics are as perplexed about the mathematics of India being algorithmic today as they were in the 1960s is a moot question.
14. Thus the renowned medieval poet and grammarian Bhartrhari's work, the *Vakyapadiya*, was not just a work on grammar but on the philosophy of language (Mahadevan 1969: 30).
15. An important point raised in these studies was that amongst the various instruments essential for obtaining valid knowledge (*pramana*) that included perception, reasoning, inference, analogy, etc., was an equally important one called *sabda*, or the testimony of a trustworthy person. It is interesting to note that in the recent literature on the sociological history of science in the Western tradition, it is recognized that the merit of a scientific argument did not merely depend on the scientific evidence mustered in its favour, but on the social station of the individual making the knowledge claim (Shapin 1994).
16. Actually, it would be an interesting exercise to figure how the structure

of Windelband's book provides a template for the structure of Dasgupta's marvellous work.

17. Nehru's two most marvellous contributions to civilizational studies, *The Discovery of India* and *Glimpses of World History*, contain very fine aperçus regarding the unity of the Indian nation, and nationhood in a broader sense. However, even though contemporary researchers on Indian history often turn to Nehru's *Discovery* for a nuanced insight, the Nehruvian corpus has been little examined from the perspective of civilization studies, save perhaps by students of international politics.

18. This author personally believes that the sting of the anti-science movements could be partially neutralized if scientists only applied the razor of critical rationality and scepticism to their own activity vis-à-vis other kinds of activity, and not just to the theories of their colleagues in the scientific community.

6

The Missing Picture*

Almost four decades after the volumes of Needham's *Science and Civilization in China* first appeared, it is fitting to ask a question that has a Needhamian twist to it: why has an Indian equivalent not been produced? In attempting to seek an answer to a question that evokes the metaphor of Cleopatra's nose, the essay proposes, rather provocatively, that the Needhamian oeuvre cannot be replicated in the Indian variant. In hoping to answer one question, we throw up a counter-factual. The proposition that a Needhamian Science and Civilization in India cannot be written may be falsified. However, the proposition could still be saved, for if *Science and Civilization* is a cultural monument that coruscated in a specific socio-political milieu, then the meanings associated with its markers, namely, science and civilization, no longer obtain in the same way.

The present essay is divided into four brief parts. The first very fleetingly outlines the history of the history of sciences in India. In the second we raise some relevant questions and problems in Needham's historiography that prove problematic for the history of science in India. In the third the conditions contributing to the non-emergence of the Indian equivalent of *Science and Civilization in China* (hereafter SCC) are discussed. A 'New Big Picture' of the history of science has emerged from sociological approaches to the history of science. What this picture holds for the history of science in India is then considered.

THE STAGES IN THE INSTITUTIONALIZATION
OF THE HISTORY OF SCIENCE

Long before Whewell's (1857) *History of the Inductive Sciences* appeared, British and French Orientalists had begun documenting the traditional knowledge forms of India within the idiom of modernity

(Colebroke 1873; Dharampal 1971; Lettres 1810). This documentation did not merely cover the mixed sciences, of interest to the Whewellian historian of science, but also consisted of detailed studies on the indigenous systems of jurisprudence, philosophical thought, including reactions of the schools of atomism (such as the Vaisesika), the evolution of logic (Nyaya), as well as comparative linguistics and philology. Naturally Orientalist scholarship prior to the *History of the Inductive Sciences* could not have been structured around the key themata that were ingrained in the history of sciences writing after Whewell's paradigmatic opus.

The history of science writing of the last three decades of the 19th century inasmuch as it addressed the non-West celebrates the emergence of modern science in Europe, almost bemoaning its absence or non-emergence in the non-West. For example, the influence of the Orientalist Ernest Renan on the French chemist Marcelin Berthelot was decisive (Roşu 1986), and the latter's *Les Origines de l'Alchimie* (Berthelot 1885), was presentist in that it hoped to recover from alchemy the traces of a positivist science, homologous to the chemistry of the turn of the century.[1] This was Whigish in the sense that alchemy was for Berthelot nothing other than a low-rate chemistry (Bensaude-Vincent cited in Besson 1992; Guillemain 1992). In any case, Berthelot himself was associated with two projects on the history of sciences related to India.

In the first instance, he and Ernest Renan were partially responsible for the inauguration of the French tradition of Indology, in particular those strands having a bearing upon the traditional medical sciences of India (Roşu 1986)—here, we have an entire lineage of scholars extending from Cordier to Filliozat (1964; 1974) (notable exceptions between their epistemological commitments notwithstanding). On the other hand the founder of the Indian school of chemistry, Prafulla Chandra Ray (1932), was ordained formally in the discipline of the history of science through his association with Berthelot. Acknowledging his indebtedness to the latter, Ray in his subsequent writing was to break away from the central axioms of Orientalist historiography of science, in the same breath laying the foundations of the social history of sciences in India.

The break with Orientalism was facilitated by a trend in economist thinking gaining ground in India during the last decades of the 19th century that sought to explain India's underdevelopment in economic terms and shelved the essentialist interpretive modalities of the Orientalists (Pandey 1992). A reflection of this strain is found in

Ray's (1918) externalist history of chemistry, as well as a large number of other writings. At about the same time, drawing upon the exemplar of the inductive sciences and a positivist conception of science, B.N. Seal (1915) was to provide the complementary internalist account in his work *The Positive Sciences of the Ancient Hindus*. Both these books must be seen as a response to the exaggerated depiction of India as a spiritual civilization, devoid of a modern scientific or industrial tradition. Therefore, it is not coincidental that these complementary works should be published around the same time as the nationalist movement began to gather momentum, acquired a cohesive ideology, and important cultural figures like Seal and Ray advocated the critical assimilation of modern science (Raina and Habib 1996).

This stage in the history of science studies in India coincides with the period when Indians commenced lobbying for the institutionalization of postgraduate teaching and research within the university system. In addition, there are independent efforts undertaken by the Indian Association for the Cultivation of Sciences devoted single-mindedly to the inauguration of a research system, followed almost three decades later by the founding of separate institutions under national control and management by the National Council of Education (ibid.). At about the same time, Jamsetji Tata moved hell and earth to found the Indian Institute of Science, Bangalore (Subbarayappa 1992) as a postgraduate research institution, entering a triangular agreement with Princely Mysore and the Government of India.

The major concerns of the history of sciences of India during this period relate to: (*a*) establishing that there existed proto-sciences, or elements of science within the geographical ambit of the Indian subcontinent based on empirical procedures, a methodology of validation, and logical reasoning (Ray 1918); (*b*) explaining the inability of these systems to develop further in terms either of the separation of theoretical knowledge from the crafts and practices, or the suppression of the sciences themselves; and (*c*) deriving a principle of quasi-continuity from the traditional sciences to modern science to facilitate the legitimation and adoption of modern science. Further, in these histories of science, science is represented as 'transcendental', and the history of science is visualized as the progress of scientific knowledge culminating in its modern incarnation.[2]

Between this stage and the next is an intermediary one, marked not so much by the institutionalization of the discipline as by the notable studies of A.N. Singh (1936a, 1936b) and B. Datta (1929, 1933) undertaken between 1927 and 1940 on the history of mathematics in India from the ancient until the medieval period. These studies belong entirely to the internalist genre of the history of mathematics. The next stage in the history of the history of sciences is marked by both a cognitive and intellectual watershed and falls in the decade beginning 1950. This is the first decade of modern science in independent India; it is also the high tide of the Nehruvian era, witness to a proliferation of industrial research laboratories under the aegis of the Council of Scientific and Industrial Research (CSIR), the emergence of the Tata Institute of Fundamental Research, Bombay, as the centre for advanced scientific research in the third world, and the inauguration of India's atomic energy programme. In this nebulous decade in which the frame of Big Science was being introduced in India, two scientific academies contended for initiating and institutionalizing the study of the history of science in India: the National Institute of Science of India, now the Indian National Science Academy, and the CSIR (Bose 1956, 1963a).

The former drew its sustenance from the initiators of the discipline in India. Needham's *Science and Civilization* inspired the programme. Both factions saw the Needhamian project as worthy of emulation in India. In the just-independent republic a new set of questions acquired currency for historians of science and, more importantly, the older questions were phrased differently. The historians in either case belonged to the scientific community, though D.M. Bose of the National Institute of Sciences had suggested involving the Asiatic Society and Bhandarkar Oriental Research Institute, Pune, for studies on ancient India. The dominant vision for the theorists of science and society during this period was Bernalist–Nehruvian. We shall outline the elements of this vision. This will be followed by a brief discussion of Needham's historiography that echoed in the projects of historians of science in India.

THE GRAND QUESTION AND THE FIRST DECENTRING

By the end of what Hobsbawm (1993) calls the 'Age of Catastrophe', a number of scientists and scientists-turned-historians of science attempted to project science as a cultural activity that enjoined humanity. This moral vision was to fill the vacuum left by the two wars. From Sarton (Pyenson 1993a) to Bernal and Needham, an attempt was under way to redefine humanism, wherein science would provide a cultural affirmative for the West ruined by two World Wars.

This 'moral' vision of science was to strike a sympathetic chord within the nebulous scientific communities in just-independent nations like India, where the bonds between state and science were mutually reinforcing (Salomon and Lebeau 1993).[3]

Within this frame science came to be characterized in three different ways (Cunningham and Williams 1993). The first is 'philosophical', where science was characterized by a method that generated knowledge structured by 'causal laws, preferably mathematical, as in the physical sciences' (ibid.). This perspective is evident in the programme of the Vienna Circle. The second characterization is a 'moral' one: science embodied the ideals of freedom and rationality, of truth and goodness, and its internalization within the heart of contemporary culture would ensure social and material progress, and possibly end all suffering and disease. The third, and possibly the most important, from the point of view of the study of science and civilizations, was the ascription to science of the quality of a cultural universal (Cunningham and Williams: 411). This meant that across all civilizations scientific activity was embodied in the 'grand universal desire to understand the world'. This activity embraced all of humanity, and its origins were traceable to the age of the prehistoric megaliths (ibid.: 408). This vision of science, deployed in several combinations of these characterizations, and best epitomized in Vannevar Bush's attempt to situate science at the centre of modern culture (Holton 1993), echoed in a multitude of nations in as many languages of those entranced by the rhetorical promise of science.

The Needhamian project proved alluring for third-world scholars since it provided a counterpoint to the Eurocentric discourse on science. Cunningham and Williams speak of the decentring of the Old Big Picture in the light of contemporary researches in the sociology of science. However, it may be proposed that the first decentring of this discourse within the Western tradition was to occur in the work of Needham.[4] Commencing with the postulate that science is a cultural universal, he proceeds to examine various 'other' ways of pursuing science in Chinese civilization. China for him is in some absolute sense the civilizational other of Europe, since India and the Arab world had been trading in goods and ideas with Europe over centuries. The China project offered him the possibility of departing from the European way of examining the history of scientific knowledge.

However, both Needham's supporters and critics have shown that he could never, naturally, entirely disentangle himself from this picture (Cohen 1994). The centrality of modern science in contemporary culture, the founding impulse of the scientific revolution of the 17th century that radically transformed European society and culture, were historical episodes that set the terrain of his investigations. Though the scientific revolution was central to the subsequent history of science, for Needham this was not part of the whole story: 'to tell this part alone is to be deeply unjust to the other civilizations. And unjust here means both untrue and unfriendly' (Needham 1973: 1). The famous river metaphor proposed by Needham (1969: 176) suggests that while there existed several streams of science that joined that of European science, the origins of modern science were in Europe.[5] The Needhamian decentring shifted the origins of science itself from Greece, and the universal script implied a more polycentric notion of science, of its emergence across civilizations.

The second decentring discussed in the last section of this chapter is a product of scepticism concerning the science as a cultural universal script, and may thereby destabilize the foundations of the formation science and civilizations studies. This also reflects the collapse of 19th-century colonial structures, emanating from Europe and mapping the world. As a result, Mendelsohn (1995: 59) writes, a train of revision of 'myths of roots and origins' is set in motion. These myths 'had become part of European self-explanation created during the late Renaissance and early modern era'. Before proceeding, we must recapitulate the idea of the scientific revolution that swept Western Europe in the 17th century, which constitutes a sort of Kuhnian hardcore around which the history of science from Whewell to Duhem to Sarton, Koyré, Butterfield, and Needham was organized. These synoptic histories, most of which were authorized during the first half of the 20th century, present the scientific revolution as one of the mind, a revolution that reorganized the geography of the imagination, and hence could only be plotted on 'maps that were metaphysical, metaphorical' (Porter and Teich 1992: 3). Further, this new science was emblematic of the age of modernity, and subsequently of industrial civilization. Naturally, historians of science were concerned in posing the question about civilizations at the threshold of the scientific revolution. Why was it that civilizations that had a significant tradition of scientific knowledge failed to generate a scientific revolution, or even an industrial revolution (Elzinga 1988)?

Needham was certainly aware of the problematic nature of his own search and the obstacles in the way of historians. This awareness is recognizable in the constant renewal of the form of the 'Grand

Question' posed by him, the first formulation of which dates back to the 1930s, when his Chinese students asked him why modern science emerged in Europe. This was subsequently reformulated to why did the scientific revolution elude China, to why didn't Chinese civilization, unlike the European, succeed in giving rise to modern science and technology, to why was there no indigenous revolution in China. In all these formulations a certain uneasiness concerning the projection of a European theory of science in history on another civilization bothers Needham. Notice the qualification in the last formulation above—he speaks of an 'indigenous revolution'. While all these formulations are situated within the framework of comparative history, the questions are re-devised as his exposure to Chinese civilizations grows. The question itself concerned what Needham (1969: 148) considered one of the 'greatest problems in the history of civilization'. Modern science and subsequently the emergence of industrial civilization appeared as major watersheds in the history of the West. For Needham, since science was a cultural universal, it was germane to the enquiry into the non-emergence of modern science in the non-West where both the state of science and techniques was comparable, and in the case of China advanced to that of Europe prior to the scientific revolution.

The later formulations of the 'Grand Question' were far more focused, unconcerned about problematic counterfactuals, and despite the inescapable schema of European history, sought to explain the phenomenon of non-emergence. The innocent question posed by his Chinese graduate students in the 1930s was subsequently rephrased as: why did modern science originate in Semitic–Occidental civilization and fail to do so in the Chinese? The more direct civilizational variant of which was: why did modern science and technology develop in Europe and not in Asia (ibid., pp. 147–54)? In more detailed studies on China we encounter the related problem: why was East-Asian culture between the 2nd century BC and the 16th century AD 'more efficient than the European West in applying human knowledge of nature to useful purposes?' (ibid. 1977a: 3–4).

Rather than address Needham's response to this family of related issues, we merely wish to focus upon the ambience generated by his ecumenical vision and the enormous possibilities that materialized in cross-cultural history, both in the West and the non-West. In order to reverse the customary tendency in the Eurocentric history of science of the ancient world to search for the shadow of Greece, he proposes a new historiographic rule. He begins by postulating that cross-cultural transmission is a natural phenomenon; and that simultaneous discovery or invention is something that the historian must establish. Since he was working with Chinese documents and inventions whose antiquity predated those of Europe, the onus of proof now resided with those who wished to maintain independent invention (Needham 1977c: xxvii, 1). Needham's sense of the constructedness of history is unmistakably evident here. Explicitly stated, if the Eurocentric account drew its potency from the idea of the Greek miracle, then a more ecumenical and inclusive history demanded the insertion of another set of historiographic rules.[6]

THE PHENOMENON OF NON-EMERGENCE

This ecumenical vision was to inspire Indian efforts in the history of science. Beyond the vision of a cross-cultural history of science, two crucial 'nodal events' from Needham's *Science and Civilization* acquire importance in studying the history of the history of science in India. But before discussing these, it is essential to highlight that Needham, to begin with, projected an European schema onto the tabula rasa of Chinese society (Cohen 1994: 450–1). This vision subsequently receded into the background as he came to be possessed by China's cultural symbolism. But this preordained vision of European science marked by the polarities of the rational and the irrational, tradition and modernity (Thapar 1993a: 25–59) provided progressivist historians of science in India with a frame to explore the non-emergence of modern science in India.

As India ambled into the age of 'Big Science' in the 1950s, three important works acquired the status of canons amongst a section of Nehruvian India's scientific community and decision-makers: Des Bernal's (1939) *The Social Functions of Science*, his later *Science in History* (ibid.: 1954), and Needham's work on China then in progress. The exemplar for the history of sciences in the 1950s was Needhamian. It is not difficult to see why this was so in the light of the foregoing discussion. Put differently, the 'high church' of science studies in India was predisposed to the Needhamian project, whilst the 'low church' drew capital from the Bernalist one.[7]

Speaking retrospectively, Bernal's *Science in History* did not offer itself as an exemplar for historians of science in India alone, but the world over. As Ravetz (1992a: 165) remarks, it was a four-volume history, the bulk of which discussed the history of the last 300 years, and made a 'disappointingly small contribution to the history of science', that it was conceptually and politically obsolete by the time it was completed. Furthermore, Bernal's exaggerated faith in the rationalism embodied in science distorted his appreciation of the science of the past. *Science in History* was drafted from a triumphalist early-20th-century perspective that could not have accommodated the non-West within its narrative fold. In short, it is at best 'externalist-Whig' history 'differing only in detail from the internalist-Whig style of historians of science dominant in his time' (ibid.: 169). For this reason, Bernal's influence was restricted to the domain of policy, a narrow section of the scientific community (Elzinga 1988), and within social movements drawn by his positivist evangelism.[8]

However, the opportunity to respond to Needham's 'Grand Question' offered the latecomers to modern science the opportunity to legitimate modern science culturally within their socio-political orders, to reinvent cognitive connections between the practice of modern science and the cultivation of traditional knowledge. In fact, it would not be an exaggeration to assert that *Science and Civilization* provided the master narrative, to be elaborated or subverted, for subsequent research into the history of science of non-Western nations.

What was the core agenda of research in the history of sciences in India as articulated by the scientific community? Why, despite the official push given to a Needhamian history of sciences in India, has such a history not emerged? We are asking after the non-emergence of a Needhamian history of science in India, even though the 'conditions' required for its production were more or less present. Floris Cohen in his massive work on the historiography of the scientific revolution, reviews the responses to the Needham question for the Islamic and Chinese civilizations. The Indian picture, among others, is missing, more or less by default. This section outlines the vocation of the missing picture.

The second leap forward in the history of sciences in India runs parallel with the creation of new institutions of scientific and technological research in the 1950s and 1960s. The National Institute of Sciences of India, now the Indian National Science Academy (INSA), organized a symposium with UNESCO in November 1951, where the idea of writing a comprehensive history of sciences in India was mooted. As far as the plan of this massive project was concerned, Needham's plan for his six volumes still to be written was considered the exemplar worthy of emulation (Bose 1956: 397–402). The most attractive historiographic principle of this scheme as reported was Needham's idea that the history of science of a region was integrated with its social, environmental, and economic history. But, overall, the plan as outlined by Bose was quite disparate, for it was unable to problematize, either in detail or in a general perspective, what the sciences in India either were like or could have been. For D.M. Bose, as for so many other Indian scientists of the time, the history of science was to provide a humanist garnishing to science, in a milieu where the rationality of science (actually technology), considered autonomous, loomed large over daily life (Bose 1956: 396-7). While commencing with the script of science as a cultural universal, the studies on the history of science that were produced under the aegis of INSA glossed over the notion of culture. The journal founded by INSA in 1966, the *IJHS*, is divided between externalist and internalist history—science in non-Western cultures is never problematized, and the enterprise, although this was not the vision of the founding fathers, is to find homologues of Western scientific theories in the Indian tradition.[9] The history of sciences of India as initiated by the National Commission on the History of Sciences and codified in the publications of INSA has developed into a journal wherein a very strict internalist–externalist dichotomy is maintained.[10] The internalism, paradoxically, is rooted in a late 19th-century positivist image of science. As internalism, an adequate epistemological account has failed to emerge.

The history of science, exceptional contributions apart, failed to acquire an ecumenical character. Bogged down by considerations of nationalist historiography, about a third of the publications in the *IJHS* addressed a range of priority disputes. The researches of the early generation of historians of science, who had grown up within the international idiom of science, made way for some revanchist writing in the post-1980s era. This new writing sought legitimacy in the existing caveats of the age: it was professedly a history free of ideological biases, and therefore was more scientific—and substituted methodological rigour and textual solidity with brave claims.

The theme of the decline of science in the non-West, be it China, India, or the Islamic world, becomes fairly dominant in the post-World War II years. Two important themes, run against each other, provide robust templates for the history of science in India. In *Science and Civilization* Needham suggests that the Mohist texts of China are the closest to the spirit of Western science. It is left to the sociology of science to answer why this school did not develop subsequently within Chinese society or why this tradition did not survive its original suppression (Needham 1977b: 171–84). For the sake of convenience, let us refer to this episode as the legend of suppression.[11] The legend of suppression is about two contending schools of thought, the one epistemically open, empirically oriented, and progressive, and the other bigoted, doctrinal, and obscurantist. The legend of suppression relates how the former is persecuted, almost eradicated by the forces representing the latter. The path of evolution of Taoism is blocked by Confucianism, the state ideology of the feudal bureaucracy (ibid. 1969: 161). The resemblance of this to the incarceration of Galileo is not lost on the readership. It may be suggested that out of the diversity of the Needham corpus, the legend of suppression provides one of the most powerful themes for Needham's Marxist following in the history of science in India.

The social history of science received a fillip outside the ambit of INSA. The efforts of two historians of science will be briefly touched upon: Debiprasad Chattopadhyaya and Abdur Rahman, both of whom drew their inspiration from Needham's work. On this count Needham's work on the Mohists and Logicians, and the Taoists was of particular importance. The legend of suppression that encapsulates the history of these schools provided Chattopadhyaya (1959, 1976) one amongst many frames for exploring the origins of materialist thought in ancient India, and the evolution of ayurvedic medicine (Chattopadhyaya 1979). Chattopadhyaya's principle thesis appears to have been that materialist schools like the Lokayata and later the ayurveda of the early period provide, possibly, the only instances that approximate to our present conception of science. The decline of these schools is imputed in part to their suppression by the religious orthodoxy. In fact, a major corpus of Chattopadhyaya's work consists in inventorying the various traditions of 'science' in ancient India, traditions that were systematically and consecutively demolished by the theocracy and those with a vested interest in the politics of irrationalism.

Abdur Rahman worked from the other end. Inspired by the twin ideas of cross-cultural history and the shifting centres of world science, he proceeded to document the practice of sciences in medieval India and in the process reveal how the medieval socio-political context and cosmos conditioned this science (Rahman 1982a, 1987). Rahman could not resist finding a counter instance to Needham's assertion that Taoism was the world's only mystical tradition that was favourably disposed to the sciences. Contrary to Needham and the findings of some Indian historians, Rahman went on to insist that the universalism of the Bhakti and Sufi movements in India could not have proved antithetical to the practice or the cultivation of the sciences (ibid. 1996).

Other than the broad historiographic issues involved, a number of sub-plots within the Needham text resurfaced in the research programmes of historians of science in India. These sub-plots were essential to the construction of the entire edifice that was an elaborate response to the Needham edifice. But this relocation, in formal terms, of Needham's findings for China in India, was underwritten by a vital assumption: similar social processes across civilizations either promote or deter the progress of science. This assumption, it may be proposed, was central to the script of science as a cultural universal, and the cross-cultural history that it later produced.

This leads us to the next question, which is a Needhamian one, and concerns the history of science in India: the non-emergence of its Needhamian variant. In a lighter vein, it may be proposed that while a Needhamian history of sciences in India has not been written, it is unlikely to be written. Nevertheless, it may be argued that the meanings associated with *Science and Civilization in China* as a cultural text between the 1950s and the 1970s no longer obtains at the end of the 20th century. And so any future rendering that may appear from Indian shores will actually be a different project despite the fact that it draws inspiration from the Needhamian one. To begin with, science and scientism no longer occupy the centre stage of human culture. But to this we should return later.

Floris Cohen (1994: 382) drops the discussion on the liminality of the scientific revolution in India on the count that 'there is as yet no historiographical tradition worth mentioning'. Cohen proposes four possible reasons why this is so. The first has to do with the absence of a scholar who could tap the misunderstood sources of indigenous science. We take this to mean that historical scholarship within the science academies was marked by scientism of the external or internal variety. The second and third are rather customary pointers to the circumstance that historical material that has survived in India does not permit such a reconstruction, or that there is no reliable chronology of Indian history. This is a much discussed shortcoming that need not be imputed to Cohen, for despite this historians have done a splendid job in India. More than anything else it reflects the inability to countenance another sense of history. Fourth, Indian civilization was visualized as metaphysically predisposed rather than scientifically (ibid.)—a fundamental Orientalist partitioning, not to mention Weber's characterization of civilizations in terms of rationality. And historians of science have challenged this characterization. Thus, out of the four historiographic points raised by Cohen, the first is the most crucial one, giving us cause to probe the sociology of the discipline in order to obtain a landscape of the social history of science in India, or to recognize the factors that curtailed its emergence.

Needham's history of *Science and Civilization* appeared against the backdrop of two political events. The first had to do with the crisis for European humanism precipitated by the two wars, and that has been referred to earlier. The second is the emergence of China in the realm of international politics in the 1940s. In a period when stable structures of democratic politics in the capitalist world came apart, 'third worldism' offered the European left the new Utopia: that the world would now be emancipated by the impoverished and agrarian periphery, who had been pressed into dependency by the 'world system' (Hobsbawm 1993: 443).[12] Despairing with what Europe had to offer in the early 1940s, three important members of the Cambridge left turned eastward for inspiration, possibly a new sense of purpose. Rephrasing a point made by Werskey (1978: 318) China was the 'functional equivalent' of what India was to Haldane and the erstwhile Soviet Union was to Bernal. The search was possibly for an alternate conceptual scheme, or theory of action, but more importantly of refashioning the science–society framework. The monumental nature of Needham's oeuvre must then be situated in the context of changing perspectives of the third world, and China in particular, in the arena of global politics.

Returning to India, prior to the attempt to give history of science in India an institutional identity in the 1950s, however unsuccessful, there already existed a fairly extensive corpus of Orientalist scholarship on the sciences of ancient India. In addition, as specified earlier, a micro-ensemble of Indian scientists had already inaugurated the 'other' picture of the history of science. These histories were drafted piecemeal as histories of specialist disciplines. They differed from each other on the finer points of historiography, and exhibited depth in areas such as the history of astronomy or alchemy. Was it possible to reconfigure this archive around a different organizing principle— for example, the Needhamian one? It more or less appears certain, if one goes by what was accomplished rather than by what was professed, that for D.M. Bose the commitment to the Needhamian framework was purely rhetorical. Further, the first generation of Indian scientist-historians of science were themselves situated within the cross-cultural perspective by virtue of their own cultural diversity, but more importantly through the influence of comparative philology that had come down through the influence of the Orientalists.[13]

In the post-war years the polarization of global politics, and the emergence of socialism and third-world societies, resulted in the canonization of Needham's work amongst the cognoscenti of the scientific world. As scientific academies and societies promoted the Needhamian project within their respective constituencies, science continued to be conceived as neutral instrumentality. The community in India, as elsewhere, was still enchanted by the two-worlds dichotomy (Snow 1959), that exhibited science's sense of cultural triumph. In third-world societies this was the golden age of scientism. Riding the wavefront of scientism, the contradictions scientism posed for a Needhamian history were possibly not evident to this generation of scientist-historians of science. Historiographically and sociologically the problem resided here.

What then of the historians? D.D. Kosambi, a mathematician-historian was in the fifties laying the foundations for drafting the Marxist history of Indian civilization (Kosambi 1985; Thapar 1993: 89–113). The history of science was not his primary concern. The formulaic application of Marxist historiography to the study of Indian civilization was first challenged by Kosambi, and it is more than likely that he might have reframed the science–society relationship had it been his primary concern. Bernal (1954) in methodologically defining science in his *Science in History* commences with Kosambi's definition of science. In the just-independent republic historians were preoccupied with redrafting the socio-economic history of India, with the study of agrarian relations, and economic historians were concerned with the potentiality of capitalism in pre-colonial India (Habib 1971).

Besides, the first generation of Indian scientist-historians of science in the early years of this century had done much to neutralize the cultural import of modern science (Raina and Habib 1996). The history of sciences of India by the mid-20th century lacked a sense of novelty and the exotic. Finally, it must be remarked that the history of science, by and large, remained throughout the 1950s to the 1970s a sort of amateur history pursued largely by scientists in their spare time or prior to retirement. Hence, there was a loss of rigour, which was possessed by the Orientalists. A major portion of the history of science writing during this period had to do with the sciences of the ancient period. Consequently, the history of science, at least for

D.M. Bose, could have been construed as a branch of antiquarian studies.[14] In historiographic terms, this history was a combination of internal–external history, inflected with empiricism and often with nationalist overtones. This framework of historiography was flat. In the mid-1970s, when alternate perceptions of science appeared, a whole new historigraphical renewal was required. A new image of science was being reckoned within social movements, and academic sociology launched a frontal criticism of science—none of these changes left a mark on the science academies that supported the history of science.

A peculiar ideological prejudice prevailed in this domain of the history of science: other than the excessive preoccupation with history of science as antiquarian studies, science more often than not came to mean the mixed or exact sciences, and to an extent the history of medicine. The history of technology remained a woefully neglected discipline, which was the bread and butter of economic historians. The concerns of the historians of science mirrored the perceptions and prejudices of the scientific community (Raina and Habib 1996). Another obstacle in the way of a Needhamian history of science has been an ideological fixation that was deeply ingrained amongst historians of science as well as general historians. Historians of science were rooted in internalist or externalist history. Historians were preoccupied with what were then perceived as more crucial questions of Indian politics and culture, but it is also likely that they were themselves partial to the two-cultures divide, and saw the externalist account as merely providing the trimmings for the internalist one.[15]

With the snowballing of developmental crises in the 1970s, complicated further by a perceived dysfunction of India's scientific institutions, a lack of engagement between the scientific and technological research system and industry, a research prioritization strategy that had little to do with the demands of India's rural population, an energy crisis acquiring menacing proportions, and so on, a revision of perspective of the science–society relationship was all but imperative. Amongst the many areas of revision, there commenced a search for alternatives, and in historical discourse science came to be problematized differently. The Indian analogue of the Kuhnian revolution coruscated from the anomalies piling up in its socio-political life. The actors who subsequently enabled the reconfiguration in science–society studies belonged not merely to the scientific community. Wearing different hats they came to be associated with social movements and the social sciences. This dimension of the story takes us away from our present problematic.

DO CIVILIZATIONAL STUDIES OF SCIENCE HAVE A FUTURE ?

The principal concern today relates to the possibility of the project of science and civilizations in a world where postmodernity denies the likelihood of a global narrative. In the postmodern world science itself becomes cultural discourse (Harding 1994), no longer a transcendent entity, but historically contingent. The script of science as a cultural universal is reduced to mere anachronism.

The history of science studies have announced the arrival of the 'New Big Picture' (hereafter NBP), discussed earlier, at variance with the 'Old Big Picture' (OBP). We shall briefly outline this picture. The NBP is a picture about contemporary society and science, but the science of the NBP is really modern science of the OBP, whose invention and not origins, since science is contingent, can be located in Hobsbawm's Age of Revolutions. These revolutions are: (*a*) the French Revolution of 1789, a political revolution that transformed the organization of society; (*b*) the industrial revolution, that commenced in Britain in 1770s, which was a revolution in the means of production, exchange, and ownership of wealth and resources; and (*c*) the post-Kantian intellectual revolution centred around the German state (Hobsbawm 1973). These revolutions enabled the emergence of a consolidated middle class that wielded political, industrial, and intellectual power (Cunningham and Williams 1993: 425). The NBP thus shifts the question of the origins of the scientific revolution in 17th-century northern Italy to the question of its invention in the later half of the 18th century. This picture has been commissioned through the efforts of sociologists of science, who in the process have deconstructed the mythology of the scientific method, and supplanted it with a pluralist principle that states that 'there are many possible ways of knowing and studying the world, and that science is just the particular way of knowing dominant in our culture'. The character of Whewellian history of science changes radically, since given the loss of science's epistemic privilege, the history of science is transmuted into a 'history of ways of knowing the world', rather than of one 'one single thing at different stages of development' (Cunnigham and Williams 1993: 409).

Rather than go into the historiography of the NBP, it would be important to fathom its impact on those drawn by Needham's 'embracing vision', if not by the nitty-gritty of his method. The NBP's principle of spatial decentring that requires treating all native knowledge forms with perfect symmetry is good anthropology of science that might assist in destabilizing essentially political arrangements claiming epistemological sanction in the name of science or modern science. But in treating modern science as the invention of the age of revolutions are we not losing out on the sense of oikumene, or limiting the history of astronomy and mathematics to the history of those disciplines in Europe over the last 300 years? Where do we situate the efforts of, for example, the lesser-known Abraham Ibn Ezra the Spaniard (d. AD 1167), considered one of the foremost transmitters of Arabic scientific knowledge to the West, and who sympathized with Ibn al-Muthanna's desire to harmonize Greek and Indian astronomy (Goldstein 1996: 9–13)? Or how do we narrativize mankind's common experience relating to the passage of the stars, those narratives that wove communities together from Toledo and Cordoba to Marageh and Jaipur? The question to be asked across civilizations is: how do particular scientific styles acquire currency and legitimacy across civilizations and nations? How does the interplay of the metaphysical, the epistemic, and the political unfold in the history of ways of knowing or a history of science? The NBP could surely find a place for these questions, but its emphasis on so recent a segment of human history provides no insight into the *durée longue* that concerns the study of civilizations. Furthermore, it appears that as far as the history of the non-West is concerned, the history of science, alias modern science, must commence with the commencement of the age of colonialism, which makes us wonder with Said (1994) whether colonialism will be a narrative that will ever be surpassed in the former colonies.

From the point of view of history of science studies in India today, a fissure continues to run right across the discipline. There is a scientist's history of science, which finds reflection even in an enlightened popular science magazine like *Resonance* published through the efforts of India's leading scientists, where society is merely an add-on (Prathap 1996). On the other hand there is a gradually growing community of professional historians of science little acquainted with the cognitive content of science. Will a new history of science in India have to sit out the enactment of Planck's hypothesis: the old generation cannot be converted, it just dies out? Arguing reflexively, the studies related to the production of theories and disciplines enable us to identify social and conceptual obstacles that could explain the non-emergence of a Needhamian history of sciences of India, but could as well lead us to question the need for a Needhamian history of sciences of India.[16] In either case, Needham's project remains an exemplar for the non-West as much as for the West. De Solla Price (1973: 9) had carefully chronicled the evolution of many a research discipline, and there was certainly an air of prophecy in a remark, possibly based on the observation of an emerging research front, that may be about the project of science and civilizations, when he wrote: 'It may well be judged that to Joseph Needham lay the honour of contributing the last great traditional and comprehensive work of human scholarship.'

NOTES

*   This paper was authored jointly with S. Irfan Habib and appeared in Raina and Habib (1999).

1.   In an interesting paper Brush (1995, 215–31) points out that while scientists are more likely to judge the history of science in terms of present-day science, a presentist predisposition to history is not identical with Whig history.

2.   The parallel with the prehistory of STS is most striking. Fuller (1993: 4) points out that during this phase, in Europe, the theorists of science and its history 'sought ways of expressing scientific claims that would move appropriately educated audiences to support emergent scientific institutions over their competitors for cognitive authority'.

3.   In the writings of Sarton, Needham, and others a non-hegemonic idea of science appears in the historical literature. Mendelsohn (1995: 63) suggests that this idea was 'overlooked by the enthusiasts of the European image of the scientific community'.

4.   One of the senses of 'de-centring' employed by Cunningham and Williams (1993: 429), drawn from Mead and Piaget was 'the process by which children come to realize ... that other people can have different visual perceptions of the same scene, and that other people can have different knowledge'.

5.   Needham (1973: 3) employs a range of metaphors to depict the circulation of scientific knowledge. Here is one conveying the spirit of the process: 'By a thousand capillary channels, like venules joining together to form a *vena cava magna*, influences come from all parts of the world.'

6.   While recognizing this possibility in Needham's oeuvre, the counterfactual history of scientific revolutions cannot entirely disentangle itself from Eurocentrism. This, it may be conjectured, has to do with the

fact that counter-factual history is enfolded within what Fuller would call an over-determinationist view of history. The view postulates certain nodal events 'through which all possible events would have had to have passed' (Fuller 1993: 210). Further, scientists reconstruct history in such a frame, paying little attention to 'contingent turning points' that reduce the number of paths available to history (ibid.: 5).

7. The high church of science, technology, and society studies, according to Fuller, includes historians, philosophers, and sociologists who seek to understand the hold that scientific knowledge exercises on society. The members of the 'low church', on the other hand, include policy-makers, political activists, schoolteachers, 'political scientists, and econo-mists—civic minded natural scientists and engineers' who have to contend with the positive or negative impact of scientific knowledge on society.

8. In fact, the radical agenda of Bernalism of the 1930s was appropriated 'by captains of industry and ministers of government in the post-war period' (Elzinga 1988: 94). Further, the science of science ideology contained prefigurations of the misconception, later implicit in prac-tice, that 'natural scientists' were seen to be 'the repository of reason and ... the most suitable for leading the world' (ibid.: 99).

9. See Chapter 5 in this volume.

10. Almost a year before the first issue of the *Indian Journal of History of Science* was launched, S.N. Sen (1966: 218), amongst the pioneers of the history of science in India wrote a piece in *Science and Culture* on the changing patterns of the history of science, where he points out the gradual shift in focus from the histories of Tannery and Comte to the external history of science. He points out that during the early stage of the history of science there is a natural tendency to accord greater emphasis to discipline histories, while the interpretive history of sci-ence must come when the historiographic tradition has reached a sufficient level of development.

11. The legend of suppression may be seen as the opposite of what Weingart (1993) calls the legend of abuse, according to which, despite suppres-sion, scientific truth prevails in the end and the scientific community stands vindicated, whereas in the former case the tradition of science declines.

12. As Hobsbawm (1993: 443) writes, theorists of the left of the first world held that 'if the world's troubles lay not in the rise of modern industrial capitalism, but in the conquest of the Third World by European colonialists in the sixteenth century, then the reversal of this historical process in the twentieth century offered the powerless revolutionaries of the First World a way out of their impotence'.

13. For a brief discussion of philology and the history of science see the chapter on Coomaraswamy in this volume.

14. D.M. Bose, Sen, and Subbarayappa authored *A Concise History of Science in India*. D.M. Bose (1963a) felt that since the Asiatic Society and the Bhandarkar Oriental Research Institute possessed expertise in antiquarian studies, they should be drawn into the project of the history of science. Sen (1966: 218), on the contrary, felt that the scientist-historian was located at a far more advantageous interpretive point than the 'non-scientist historian' who 'may not only overlook the significance of a scientific work, but is more likely to misinterpret it'.

15. See Elzinga 1999, where he discusses Shapin's account of the asymmetric relation between the internal and external history of science. While the former could have an independent existence, the latter is parasitic upon the internalist account.

16. In a recent review of two books on the history of sciences in India Chris Bayly (1996: 33) raises the inadequacy this paper has sought to address. Bayly writes: 'While there are many worthy tomes and articles on Indian learning, it still lacks its Joseph Needham to synthesize its mass of findings and project its importance to the forefront of international scholarship'. When posed this way, we think the onus for the production of this Needhamian history must lie with the West. For as in the case of China, so in India, as this paper indicates, there existed a corpus on the history of sciences prior to the arrival of the Needhamian exemplar.

7

Reconfiguring the Centre

Studies of the introduction of the traditions and institutions of modern science—or the exchange of scientific knowledge between the metropolises of modern science and the 'periphery'—have typically addressed the transmission of scientific knowledge from the former to the latter. In addition, there have been many detailed studies of the growth of science through scientific expeditions undertaken at the periphery by scientists from the centre.[1] Most investigations of the extension of the dominion of science within the 'centre and periphery' model have used the familiar exemplar of shifting centres of science in 18th-century Europe: from Britain to France to Germany (Gizycki 1973; Nakayama 1991; Shils 1991), and from Europe to the United States (Schott 1993).

The 'centre–periphery' framework, it may be suggested, rests upon the twin propositions that science is a cultural universal, and that its reception by the periphery is unproblematic. However, encounters between leading European scientists and leading Indian scientists illustrate some of the problems of employing the centre–periphery model. Biographies of scientists born at the periphery fall into a pattern that invites comparison with origin myths' (Viswanathan 1992a) and legends of abuse (Weingart 1993). The local scientist is an exile astride two worlds—the butt of ridicule among competing countrymen and an exotic specimen to colonial administrators. The colonial subject is regarded by the colonizer as receptive to the civilizing mission and can only be recognized by the practised eye of the expert.

In India, and elsewhere in the colonial world, scientific claims emerging from the periphery were least problematic once scientists from the periphery gained entry into the 'invisible colleges' at the centre. Towards the end of the 19th century and the first decades of this century, India was at the periphery of modern science. While 'collegial ties' and 'cosmopolitan ties' were forged across 'social distances and long distances', (Schott 1993b: 198), successive encounters between Yesudas Ramchandra and Augustus De Morgan, Srinivasa Ramanujan and G.H. Hardy, P.C. Ray and Marcelin Berthelot, and M.N. Saha and his astrophysicist colleagues in the United States reveal much about the ways in which 'the contemporary global institutional frame shape[d] the communal participation in science'(Schott 1993a: 197).

LEVELS IN THE 'CENTRE–PERIPHERY' MODEL

While the model provides a broad framework for observing the exchange of scientific knowledge, the pertinent question is how it affects specific research programmes. For example, in the 1920s, Saha's encounter with the American astronomer Russel—both of them were then at the periphery of the scientific world—simultaneously created the conditions for the emergence of a new discipline, despite their location on the 'periphery'.

It is useful to consider this question within a two-tier, or two-level, structure. Confining ourselves to the 19th century, at Level One are interactions between scientists at centres and peripheries within Europe; these are structured by particular sets of relations, which define the practice of science. Level Two concerns interactions between European centres and the periphery of science outside Europe (Ben-David 1984). These two levels encompass exchanges at the macro level. But what is the nature of exchanges within specific research programmes?

Exchanges at different levels may be similar in structure but different in dynamics. For example, there may be similarities between centre–periphery relationships among scientific communities within Europe (Level One), and relationships between the latter and scientists at the peripheries outside Europe (Level Two). The propositions essential to Level One have also been applied to the study of non-European contexts. This argument is predicated on two propositions. The first, following Gizycki, states that the centre–periphery framework characterizes a hierarchy in the production, distribution, and organization of scientific knowledge. Thus, research undertaken at the centre 'commands more attention and acknowledgement than works produced elsewhere; a centre is the place from which influence radiates'. Such centres are not all encompassing but discipline specific (Gizycki 1973: 474).

If these are the defining characteristics of a centre, Schott's discussion of the globalization of science specifies the collegial properties of scientists working there: homophilly, homogeneity, density, multiplexity, and range (Schott 1993b: 201).[2] The global scientific community incorporates both levels in such a way that globality, hierarchy, and division of labour organize the world of science into a homogeneous web (ibid.: 204). Hierarchy is further 'black-boxed' by relationships of dependency, and a model of exchange considered as a unidirectional transmission of ideas. If one changes the focus of investigation from 'transmission' to either 'redefinition', 'exchange', or 'translation' as construed by Latour (1993: 7–16: 253), the hitherto passive recipients are transformed into active agents.

The second proposition relates to Level Two. Within the history of science the centre–periphery model interlocks with the research themes 'science and empire' and 'science and colonialism'. The corresponding knowledge form is referred to as 'colonial science' (Basalla 1967). The proposition asserts that colonialism produced forms of social organization, which in turn set the coordinates for the production of scientific knowledge in colonies.

The hierarchies in the production of scientific knowledge constitute a political field, so investigation at Level Two should be sensitive to the politics of knowledge as well as to the manner in which science is instituted as a cultural universal (Polanco 1985). One way of prising open the black box is to look not merely at flows of knowledge, but also at the processes whereby knowledge is exchanged.[3]

At the turn of the 19th century what values were shared by scientists at the two levels? These values can be called epistemological and axiological. Epistemologically, it could be said that scientific ideas and theories were validated in a pristine space, their merit being decided in the court of reason and evidence. To its public, science was presented as a 'theatre of proof' (Inkster 1975; Thackray 1979; Salomon-Bayet 1986). In axiological terms, the culture of science was international, and internationalism was a prerequisite for the advancement of scientific knowledge.

In the late 19th century Louis Pasteur embodied the interplay between the national prestige inherent in scientific achievement and the internationalism of science (Dubos 1950: 85). Pasteurian internationalism consisted of several elements: the model scientist was a citizen of the world, not of any one country; the procedures and results of scientific research were valid irrespective of the nationality of its producers; scientists were concerned about the contributions of their countries to the international stock of knowledge, but this 'did not infringe upon the solidarity of the scientific community'; and whatever the 'deficiencies in the actual conduct of scientists, [it was assumed that] they should behave in accordance with the universally valid, common substantive and procedural traditions of science' (Gizycki 1973: 478).

The last can be seen as a central tenet of internationalism. For a scientist coming from a European colony, the normative values of science at the centre operated once entry had been granted into the invisible colleges. While the epistemological account provided a stable frame of reference, the rules of admittance into these collegial circles varied considerably between different colonies and metropolitan centres.[4]

COLLEGIAL RELATIONS UNDER A COLONIAL REGIME

The problem of analysing colonial relations within specific research programmes is conventionally posed in terms of metropolitan influence on colonial institutions. However, they also embody forms of patronage to be studied in socio-epistemological terms. The nature of the centre–periphery exchange has to be considered in the light of developments within local scientific institutions and their role in the professionalization of science.

Between 1840 and 1920 India experienced a gradual change in the nature of scientific activity, as the amateur and autodidact were gradually replaced by the professional scientist. The mathematician and journalist Yesudas Ramchandra may be taken as a representative figure. Later, there emerged scientists who may be seen as cultural analogues of the German *Kulturträger*. The chemist P.C. Ray, physicist J.C. Bose, and founder of the Indian Association for the Cultivation of Science M.L. Sircar belong to this genre. While some of these scientists were renaissance men, their efforts were instrumental in inaugurating a third phase, that of the professional scientist. By the 1890s most of the large presidency towns in India had universities, but research was not in their charter. The task of the leading members of the Indian Association for the Cultivation of Science was to found a scientific research system. The year 1914, when the First Indian Science Congress was organized, was a landmark in the professionalization of Indian science. The research system thus emerged in tandem with the nationalist struggle (Viswanathan 1985). The encounters between Ramchandra and De Morgan between 1845 and 1870, and between Ramanujan and Hardy between 1900 and 1917 illustrate the interactive processes by which the agencies of colonizer and colonized were expressed. While these exchanges took place within Level Two, the scientists involved were informed by what occurred at Level One, which in turn shaped their desire to reconfigure Level Two.

Ramchandra and De Morgan

Yesudas Ramchandra was a teacher of science, a man of letters, amongst the founders of modern journalism in Urdu, and a progenitor of popular science writing in Urdu (Habib and Raina 1992). He spent the 1840s introducing calculus to Indian students, and in so doing confronted serious pedagogic problems, then current in Europe as well as in India. Ramchandra's experience illuminated the conditions governing the acculturation of modern mathematics in a non-European context.

In the encounter between Ramchandra and De Morgan, patronage played an important part. Some forms of patronage are specific to scientific activity, and each form seeks to solve the 'uncheckability problem' that arises from the uneven distribution of knowledge and discretionary power (Turner 1990: 187). Stephen Turner has proposed a list of devices employed in establishing trust between scientists through patronage: attestation, metonymy, and extra-scientific relations. None of these works to the exclusion of the others, but in any given situation combine in establishing trust (ibid.: 193). The inflection they acquire within the patronage system of science, against the backdrop of colonialism, becomes easily apparent in the asymmetrical distribution of knowledge and power between centre and periphery.

Ramchandra was schooled in the 1830s in the algorithmic tradition of mathematics in India, and was equally at home with the mathematics in school curricula established by the British. His pedagogical uneasiness relating to Euclidean geometry, as taught in Britain, first found expression in a mathematical work written in Urdu for schoolchildren, called the *Sari-ul-Fahm*. His *Treatise on the Problems of Maxima and Minima*, published in India in 1850 and in London in 1859, could in the light of contemporary mathematical knowledge be seen as a non-topological attempt to introduce elementary calculus within a culture where Euclidean geometry was not highly valued.[5]

The *Treatise* was in English, and in the tradition of 19th-century textbooks of mathematics. However, it contained a new method designed to solve what Augustus De Morgan, professor of mathematics at University College, London, later referred to as 'Ramchandra's problem'. De Morgan noted that Ramchandra could have published his findings in a research journal, but his purpose was primarily pedagogic. De Morgan was among the first established mathematicians to see a copy of Ramchandra's work. At Level Two, for Ramchandra, England was the centre. On the other hand, within Level Two, Ramchandra's work was severely criticized in India for its lack of novelty.

Indian scientists were ignorant of the rules of the British scientific community that was itself undergoing rapid professionalization. They had to be tutored in the rules and etiquette of the 'collegial circle', the 'extensive communality' that characterized the world of science (Schott 1993a: 196). Ramchandra and De Morgan occupied very different posts within their respective educational systems, the former being a schoolteacher and the latter a university professor. The space available to forge such a relationship of trust was that of a shared problem—the place of algebra in a liberal education (Raina 1992; Raina and Habib 1990)—and a shared perception of how the history of mathematics could inform contemporary mathematics education.

This could be the rationale for De Morgan's elaborate preface to the London edition of Ramchandra's *Treatise*. It may be argued that a preface or a foreword is a form of attestation, and that De Morgan's excursion into the history of the mathematics of India is a sign of the hierarchies of relevance operating within the domain of science. The problem for De Morgan was how to explain to an audience tutored in the English mathematics curriculum of the 1850s that mathematics could also be done in other ways (Hodgkin 1986). The quaint mathematics from the periphery had to be decoded through an established interlocutor in order to make it relevant to mathematics as practised at the centre. From the centre's point of view, it was a way of increasing the 'range' of the collegial circle (Schott 1993: 201).

In terms of centre and periphery, the experience reveals a number of features. First, the exchange was at Level Two and conformed to the traditional idea of how scientific knowledge is transmitted, with Ramchandra seen both as 'native informer' and the *indigene* who had undertaken the task of translating modern calculus into the Indian idiom. But Ramchandra's *Treatise*, produced on the periphery, also had importance for De Morgan at the centre. De Morgan was engaged in establishing a place for algebra in a British liberal education (Pycior 1983), and suggested that the book be republished for circulation in British schools (Raina and Habib 1990). If we reconceive the process of 'translation' as defined by Latour, then the centre was not just the source of influence, but the site where this attenuated knowledge from the periphery was reappropriated. The similarities in the accounts of the lives of Ramchandra and Ramanujan are surprising, although there can be no comparison at the level of mathematics.

Ramanujan and Hardy

The relationship between Ramanujan and Hardy has been of interest ever since Ramanujan scholarship acquired its enigmatic place within the history and sociology of science (Shils 1991). Would Ramanujan have been a genius had he been tutored by Hardy and Littlewood into the idiom of turn-of-the-century Cambridge mathematicians? Had Ramanujan not gone to Cambridge, would he have founded a school of mathematics in India? And why did he turn to Cambridge and not to France or Germany?

These questions suggest that all paths lead inexorably to the centre; the centre is like an invisible hand controlling the trajectory of scientists at the periphery towards the fount of wisdom. Thomas Schott (1993: 448) points out that research from the centre diffuses easily, either because it emanates from the centre and hence is valued over and above its intrinsic value, or because it influences the choice of research and methods at the periphery. The scientist-mathematician schooled at the periphery is at a disadvantage regarding problems that are considered worthwhile within the scientific community at the centre. Research problems of current interest cluster within a range of theoretical and empirical possibilities, and are expressed in a language of consensus seeking accepted by the community (Shapin and Schaffer 1985: 25). In 1910, Britain was no longer a mathematical centre of first importance, the international centre having moved to France and Germany. But within the Anglophone world—the world within which most colonial subjects in the British empire had regular access—Trinity College, Cambridge, remained a centre.

The relationship between Ramanujan, Hardy, and Littlewood, at the high tide of colonialism, stands out as an exemplar to those who swear by the axiological autonomy of science—in epistemological terms, science leaps forward independently of colonial prejudice.[6] That account is, nevertheless, asymmetrically oriented towards the centre. The centre–periphery relationship, as visualized at the centre, may be a symbiotic one, but it may also prove parasitic for the periphery. Ramanujan's move to Cambridge strengthened Hardy's research programme on the formalization of number theory. In India Ramanujan was to become a source of inspiration for subsequent generations of number theorists, but 'to Indians at the time (1915–30), Ramanujan was not unique in the way we think of him today. He was one of others who had, during the same period, achieved, in their judgement, comparably in science and in other areas of human activity. Gandhi, Motilal and Jawaharlal Nehru, Rabindranath Tagore, J.C. Bose, M.N. Saha, S.N. Bose, and a host of others were in the forefront of the then fermenting Indian scene' (Chandrasekhar 1987).

Apart from theorems sent by post, the correspondence between Hardy and Ramanujan reflects the hierarchies operating within the world of science and also the structure of patronage within the colonial world. Robert Kanigel (1991: 159), in his biography of Ramanujan, describes the rhetoric of supplication that Ramanujan adopted in appealing to Hardy to give him a reasonable hearing— introducing himself in a letter to Hardy dated 16 January 1913 as 'some insignificant clerk in some backwater of an office five thousand miles away (who) apparently sought to incite both pity and wonder'. The normal opportunities for collegiality were not available to Ramanujan, who wrote: 'I have not trodden through the conventional regular course which is followed in a University course.'

C.P. Snow described Hardy's initial reaction on receiving Ramanujan's letter and theorems: 'Hardy was not only bored but irritated. It seemed like a curious kind of fraud.'[7] Later the same day, however, Littlewood and Hardy decided that they had 'discovered [a] natural mathematical genius' (Hardy 1992: 33). Retrospective narratives describe the forging of trust by emphasizing the 'adventitious and serendipitous' aspects of the encounter as 'central to the discovery' (Turner 1990: 190). The discovery is represented as that of a jewel lost amidst chaff. As this account suggests, Hardy's trust in Ramanujan was established when Ramanujan joined him at Cambridge. The establishment of trust dispelled the need for trust (Kripke 1982: 69).

A matter of contention arising from the discovery of Ramanujan's mathematics—one that Shils recognized but interpreted differently, seeing science as a privileged way of knowing—relates to how a mathematician at the periphery could invent a mathematics that surpassed insights available at the centre. This is another way of asking how the framework of centre and periphery handles such temporary disruptions in flows of influence, which is probably why Shils found Ramanujan's mathematics to be problematic within that framework. How could such reversals be incorporated in the framework? A dynamic centre is not only a source from which influence radiates, but also one that readily receives the voices that 'surpass it'.

A central postulate of the centre–periphery framework—that the centre is a privileged site for the generation of influence—needs qualification to account, for example, for the contributions of S.N. Bose in Calcutta to the formulation of quantum statistics in the 1920s, and of M.N. Saha, also in Calcutta, to the incipient programme of theoretical astrophysics and astronomy.[8] It must also account for C.V. Raman's contribution to the phenomenon of scattering, Heidekei Yukawa's development of meson theory, and Tomanaga's contribution to the study of quantum electrodynamics. The pattern of emulation set up by the centre triggers efforts at the periphery that may surpass the centre. And while the centre offers a model for the periphery, the latter nourishes the global scientific community through its own social and cultural practices (Elzinga 1993: 141).

The centre construed as the fount of influence requires that the periphery be seen as a source of data. The periphery—given the prevalent hierarchies within science—is then not a site from which theoretical frameworks emerge or new theoretical systems can be enunciated. The only way the model can overcome these limitations is to revise the transmission model upon which it rests. This revision is suggested by historical contingency—'the centre always shifts' (Nakayama 1991).

BETWEEN NATIONALISM AND INTERNATIONALISM

When internationalism is confronted by nationalism, relationships at Level Two enter a new dimension.

P.C. Ray and Marcelin Berthelot

The exchange between P.C. Ray, the founder of the school of synthetic organic chemistry in India, and the eminent French chemist, Marcelin Berthelot, is markedly different from the experiences of Ramchandra and Ramanujan. First, Ray was trained as a chemist and obtained a D.Sc. from the University of Edinburgh in 1888, under the supervision of Crum Brown. Ray was familiar with the research programmes of Brown and Thomas Fraser at the conjunction of synthetic organic chemistry and pharmacology. Second, Ray was tutored at one of the centres, and hence bypassed the stages of attestation to which his forerunners had been subjected. In 1889 he returned to India and worked towards the establishment of a pharmaceutical industry and a school of research in modern chemistry.

Ray set out to emulate the centre at the periphery, but this exercise proved impossible to carry out in detail. In any case, his inspiration was not Edinburgh but Germany, the reigning centre of organic chemistry (Ray 1932). The task was to carry over the exemplar of the centre from Level One to Level Two.

While Ray was from the outset a member of the network of leading chemists, both Ramchandra and Ramanujan were autodidacts, outside the collegial circle of their disciplinary field, though their work was reviewed and assessed within their disciplines. The subtle tension in the early letters between Ramanujan and Hardy indicates how much scientific communication relied upon conventional norms and forms of patronage. Ramanujan exasperated Hardy by seeming to pull formulas out of a hat. Hardy wrote to him: 'I want particularly to see your proofs here ... in this theory, everything depends on the rigorous exactitude of proof ... depend(s) on absolute rigour' (Kanigel 1991: 173). Ray on the other hand operated within the same sphere as his peers; he was vulnerable since he was contesting the claims of science emanating from the centre, while he himself was at the periphery.

In centre–periphery relationships there exist a sphere of patronage and a sphere of competition. The relationships between Ramchandra and De Morgan and between Hardy and Ramanujan reveal different aspects of patronage, and show how they structured the exchange of knowledge. Ray, by graduating from Edinburgh, did not have to 'establish' trust. His position was further validated by the fact that he held a position in a university department of chemistry. From the realm of patronage he entered the sphere of competition.

Within this sphere, the rules were those of reason and evidence, and the promise of benefits that science would bring through the internationalism of ideas. Ray was a member of a disciplinary research community considered to be relatively autonomous vis-à-vis external pressures to perform relevant research (Elzinga 1993: 144). The centre–periphery framework does not allow for the possibility that, even within such communities, there may be 'epistemic drift'.

The model is founded on the presumption that the exchange of knowledge between centre and periphery is guided by purely epistemic criteria, whereas other 'external' factors may be present (ibid.).

In Ray's exchange with Berthelot, the field of contest was not so much scientific theory, but the history of science, which in the late 19th century was a narrative of progress and development, overlaid with the presuppositions of nationalist historiography (Adas 1990). Thus within the realm of scientific exchange arose a recognizable 'double contingency (Elzinga 1993a: 144), relating simultaneously to science and politics.

On his return to Calcutta in 1888 Ray began to investigate the materia medica of India. In the process he encountered Berthelot's *History of Greek Alchemy*. Berthelot's interest in the alchemy of India inspired Ray to begin a historical investigation of Hindu alchemy, and a correspondence ensured. Ray believed there was insufficient evidence to support Berthelot's thesis that the origins of Indian alchemy were traceable to Greek sources, and held that Chinese influences could not be ruled out. Nevertheless, Berthelot was an iconic figure in the world of chemistry, particularly for young Ray who had committed himself to the establishment of a school of industrial chemistry, which required the establishment of an industrial research system. Ray's study of the origins of Indian alchemy was to conflict with Berthelot's authority.[9]

The important feature of this encounter is the tropology characterizing an exchange in which two different views of history were at work. With Berthelot's encouragement, Ray discovered that the use of mercury and mercury-based compounds, and the preparation of several caustic compounds in the medical traditions of India predated Arab influence. Berthelot argued that this knowledge was transmitted from Greece to Egypt and then to the Arab world via the Syrian Nestorians, and from there to India and China. This disagreement, which arose from distinct views of history, Berthelot ended by accusing Ray of nationalist prejudice. Within the transitional community of scientists, where internationalism was the norm, the term 'nationalist' was used pejoratively. The idiom of science had been formalized through the prevalent 'literary and scientific technologies' and knowledge arising from the periphery could be discredited from the centre by evoking the violation of the tenets of internationalism.[10] Political theorists recognize that within the colonial world the constitution of the periphery as the 'empire of unreason' prompted reverse commentaries from the periphery (Kaviraj 1988b: 12–13). The exchange between Berthelot and Ray may be regarded as a classic case where the 'understudy' spoke up.

Ray put his case in the court of reason, and Berthelot had to accept Ray's argument without surrendering his own authority. The disputed issue was pushed into the background, and in the second review of Ray's book, published in the *Journal des Savants* in 1903, Berthelot concluded with the lines: 'An important chapter had been added to the history of the sciences and the human spirit, a chapter that was particularly useful in its understanding of the reciprocal relations that existed between the Oriental and the Occidental civilizations' (See chapter on P.C. Ray in this volume). The conventions of the international community of scientists structured the interaction between the two chemists, once Ray had been granted entry into the community of practitioners at the centre.

Norms may be diluted or changed in the process of resolving controversies, particularly if their subject matter impinges upon the central perceptions of the communities concerned. But this typically ensured the hegemony of the centre. The realization of Berthelot's project depended upon the successful integration of a detailed and thorough history of alchemy, since the work was not merely about the origins of chemistry, but also about the history of scientific progress. Berthelot sought to chronicle the escape of modern chemistry from the clutches of alchemy, and thus his work was intended as a panegyric to the victory of reason. For Ray, Berthelot, whom Ray made a point of meeting, continued to remain a role model. In 1907 Ray dedicated the second volume of his *History of Hindu Chemistry* to the memory of Berthelot. From the periphery, the centre still held the place of honour.

M.N. Saha and the Birth of a Discipline

In contrast, we have the experience of M.N. Saha, a leading Indian physicist and the founder of the Saha Institute of Nuclear Physics. A student of Ray, Saha was awarded a D.Sc. in 1919 by Calcutta University. He belonged to the generation of Indian scientists who graduated at Indian universities under the influence of scientists trained at the centre, and who were committed to the construction of a research system that would be independent of the centre's patronage. The programme acquired increasing stridency in India as the political struggle for freedom from British rule gathered momentum (Abrol 1995).

What happened when Saha, a scientist at the periphery, competed for a position of leadership in the metropolis? The question reveals a transformation in the colonial patronage relationship. DeVorkin and Kenat have identified two phases in the history of astronomical spectroscopy in the first three decades of the 20th century: a phase of qualitative chemical analyses, and a phase dealing with quantitative chemical analysis and the structure of the solar atmosphere. The principal problems during the period of transition from the qualitative to the quantitative eras of astronomical spectroscopy related to the detection of line spectra and atomic structure, the determination of chemical abundances in the solar and stellar atmospheres, and the role of temperature and pressure in the stellar atmosphere in producing differences observed in the spectra of stars (DeVorkin and Kenat 1983a: 103, 1983b). The Hertzsprung–Russel diagram and the technique of spectroscopic parallaxes facilitated the interpretation of stellar spectra. This was necessary for deciphering the relation between stellar temperatures and luminosities.

Saha came to the problem through teaching thermodynamics and spectroscopy at Calcutta University, where he 'devoured all available issues of European and British scientific journals' (ibid.: 111). The principal influences upon his work were the publications of Bohr, Nernst, Sommerfeld, and Eddington. The ionization formula derived by Saha revealed for the first time the interrelationship of the total pressure of a gas with the degree of ionization of an element in the gas, the ionization energy of the gas, and the temperature of the gas. This theory was worked out in the early 1920s through a series of papers, which provided a rational theory of spectra, reorienting the efforts of two schools of astronomical spectroscopy, one at Cambridge and the other of Russel at Harvard. Russel anticipated the application of Saha's theory and 'altered significantly his own research program to exploit it' (ibid.: 126). The theory was rapidly accepted among astronomers who saw the possibility of application. Russel tinkered with Saha's assumptions and refined the equation to suit 'a more realistic situation' (DeVorkin 1994: 158).

Thus, the question again arises: how could science at the periphery surpass science at the centre? DeVorkin (ibid.: 126) suggests that Saha's 'relative freedom in isolation' enabled him to tread entirely new pathways, although it constrained him from exploiting the potential of his theory. This suggests that the scientist at the periphery was unaware of the full potential of his formulations since these were determined by what the centre considered relevant.

Saha's fate, it is said, was sealed by a 'weak and divisive infrastructure for science in India' (ibid.: 157). But, while the prospect of founding a school in India was limited during the colonial period, this interpretation shifts the responsibility for Saha's failure away from the asymmetry that typically characterizes the relationship of centre and periphery. There was a similarity in the positions of Saha at Level Two and Russel at Level One. Both were peripheral to physics—'Saha geographically, and Russel professionally'—and both were thereby free to explore various niches for the application of the theory (ibid.: 158). Both also 'possessed great synthetic powers', which enabled them to apply a range of tools to 'broad classes of evidence pragmatically to reach provocative conclusions' (ibid.: 158).

Saha's efforts in India confirm the thesis that the possibility of idea hybridization at the periphery is greater because the pressure of conformity within institutions, academies, and disciplines is lower—thereby compensating for 'conformist pressures of modern science' (Chayut 1994: 298). The hypothesis that hybrid roles redirect the efforts of scientists towards new problems has substantial historical precedent. Nevertheless, to a large extent the marginality of science at Level Two restricts the development of a new community at the periphery. Possibilities are limited by the lack of well-endowed and well-regarded centres.

CONCLUSIONS

The 'centre–periphery' relationship historically structured scientific exchanges between metropolis and province, between the fount of empire and its outposts. But the exchange, if regarded merely as a one-way flow of scientific information, ignores both the politics of knowledge and the nature of its appropriation. Arguably, imperial structures do not entirely determine scientific practices and the exchange of knowledge. Several factors neutralize the over-determining influence of politics—and possibly also the normative values of science—on scientific practice.

In examining these four examples of Indian scientists in encounters with their peers at the centre, exceptional scientists are seen in a social context where the epistemology of science supposedly describes its practice. Imperialism imposes practices and patronage, which moderate the exchange of scientific knowledge. But at Level Two the politics of knowledge and the patterns of patronage within it mediate exchanges between the centre and the periphery.

The first step in reconfiguring exchanges between centre and periphery—in this case, between Europe and India during the period

1850 to 1930—is to recognize the relation between the acquisition of resources and the maintenance of legitimacy and identity.[11] Political life is not confined to the core of political institutions (Hasse et al. 1995). Second, in examining science as practised in the colonies, it is necessary to see stages of scientific institutions whose development structures the exchange.

From the encounter of Ramchandra and De Morgan, it is evident that the centre–periphery framework should be separated from the models of transmission embedded within it. The notion of 'translation' helps to suggest that scientists bring personal motives and meanings to each encounter. Ramchandra, for example, sought a novel method of teaching Indians calculus, while De Morgan's interest lay in finding a place for algebra in a liberal education.

The hierarchy inherent in the centre–periphery framework compels the conclusion that at Level Two the autodidact outside the institutions of science must have his work presented to scientists at the centre by authoritative figures from the centre. This is not mainly a question of imperialism, but rather of patronage. The peripheral scientist could not be granted direct entry into the collegial circle until his efforts at the periphery could be translated into the language and concerns of the central community. Ramanujan's enigmatic formulas were translated into the language of analysis by Hardy, which enabled the creation of a field to which Hardy was committed.

Scientists from the periphery who were already part of the circle by virtue of their training were not necessarily subject to the same degree of attestation as other scientists from the periphery. P.C. Ray, with his D.Sc. from Edinburgh and his position at Calcutta University, had less difficulty in winning the trust of colleagues at the centre, even when he returned to India. On the contrary, remaining at the periphery, he moved from a context of patronage to a sphere of competition. In addition, Ray's collegiality, even at Level Two, was more comprehensive, and connected him with Level One.

Eventually, the professional Indian science graduate found collegiality within the international community of scientists. Saha's self-imposed progressive nationalism constrained his identification with the centre and made him a potential competitor instead. Once having achieved eminence in the world of science, C.V. Raman and Saha shifted their work to journals of physics published in India in order to further the cause of physics research in their own country (Raina et al. 1995).

To go beyond the limitations of the centre–periphery model, it is necessary not merely to examine exchanges between scientists functioning in a 'shared epistemological universe' (Sörlin 1992), but also to recognize the part played by institutions, the experience of colonialism, and the forms of patronage characterizing both colonialism and science. Put another way, although there is relative epistemological autonomy within the disciplinary research communities of science, the interplay between knowledge and power structures this exchange.

The scientific links between colonial India and Britain at the turn of the century were mediated by structures that prefigured change. Does structure determine all? If it does, we are left with an Orientalist reconstruction of the docile native and a passive cultural medium into which science percolates. But this neglects the role of scientists in creating new structures within which they worked. A middle position—one more sensitive to the exigencies of colonial scientific life—would be one where the participants are seen not as the dupes of 'structure nor the potentates of action', but as occupying a ground between the two (Cozzens and Gieryn 1990: 14; Hagendijk 1990).

NOTES

1. A work in this genre is Pyenson (1993a).
2. The centre has been defined diversely. Macleod (1987) considers metropolitan science to be a way of doing science based on learned societies, cultivators, conventions of discourse, and theoretical priorities that were set in 18th-century Western Europe.
3. Recent sociological critiques of the transmission paradigm have revised our appreciation in that the cultural consumption of scientific knowledge is tantamount to cultural production. See Cooter and Pumfrey (1994).
4. Put differently, 'The doctrine of progress as global civilization promoted a cosmopolitan orientation among participants. Natural inquiry was to be open to participation from any part of mankind (only later woman), and the resulting knowledge was to be widely disseminated as a collective good of humanity' (Schott 1993b: 198).
5. I shall not discuss here whether geometrical notions are extant within the Indian tradition. See Raina and Habib (1990); Habib and Raina (1992).
6. Hardy's (1940, 1967: reprinted in 1992) autobiography, *A Mathematician's Apology*, could be read as a text on the imperial mentality. Kanigel (1991: 171) claims that this knowledge of India could have been 'mired

imperial stereotypes', without Ramanujan being tainted in Hardy's eyes.

7. C.P. Snow's foreword to Hardy's (1967: 31) autobiography.
8. Saha, following the recognition of his work on the ionization formula, wrote to the American astronomer Hale requesting support to pursue the programme he was developing. However, Hale and Russel were engaged in their spectral investigations at Mount Wilson. Hale informed Saha that they were following the agenda as proposed by him, but 'Saha was not invited by his European and American colleagues to collaborate with them in refining and extending the theory' (DeVorkin 1989: 98).
9. See Chapter 3 in this volume.
10. Rephrasing Shapin and Schaffer (1985: 25), the literary technology here would refer to how the content of the subject would be presented to the non-expert, while the social technology would incorporate conventions scientists used in relating to each other's claims.
11. For a discussion of the neo-institutionalist approach in science studies see Hasse et al. (1995).

8

From West to Non-West

The diffusion of modern science to the 'new' and 'ancient' worlds from Europe in the 19th century constitutes a rapidly developing area of research in the history of science. Depending on the discipline, the focus of investigation is either the transmission of modern science and technology to the non-West or the globalization of modern scientific practice (Headrick 1981, 1988; Pyenson 1985, 1993b; Schott 1993a, 1993b). These studies are characterized by a diversity of approaches. The early studies assumed science as a positivist given, and thereby attempted to trace the unattenuated epistemological transmission of scientific knowledge from the sites of its production in Europe to the domain of ignorance. Studies cognizant of the political entanglements of scientific knowledge reckoned with the circumstance that science in the process of its transmission served as an instrument of cultural imperialism and colonial expansion (Pyenson 1993a; Headrick 1988). Drawing from economic theories emanating from the developing world, sociologists of science postulated the notion of the scientific centre as the fount of influence that radiated to the relatively underdeveloped periphery (Ben-David 1984; Gizycki 1973; Nakayama 1991), and elaborated another framework for the study of transmission of modern scientific knowledge. On the other hand a strain of scholarship rooted in the comparative sociology of science and history highlighted the contingent character of the scientific revolution in the West (Needham 1977a; Nelson 1981; Zilsel 1942). This realization prompted interest in the institutional structures that generated scientific knowledge and that have subsequently enabled the growth of modern science (Ben-David 1984; Merton 1970).

This essay focuses primarily upon those researches addressing European expansion and the expanding dominion of modern science.

These accounts, which have largely been linear in form (Storey 1996: xiv–xv), refuse to reckon with indigenous conceptions of the natural world, and the dialectical interplay between modern scientific knowledge and traditional theories of the natural world. In fact, these studies have restricted their historical concern to the implantation of modern scientific knowledge and the institutions of modern science in non-Western environments. In other words, these studies embody a non-interactive approach, for they merely view the non-West as laboratories for the performance of scientific experiments.

Interactive approaches to the history of transmissions have arisen from developments in the sociology of scientific knowledge and the comparative sociology of science (Huff 1993: 33–46). With the erosion of the ideal of the value neutrality of science and the demise of positivist theories of scientific growth, linear models of the extension of the dominion of science came to be challenged. Furthermore, the natural evolution of the history of science, driven by a universalist impulse paradoxically challenged Eurocentrism. This change was not merely a result of the internal development of the discipline, but stimulated as much by the rise of socialism in the 1940s and 1950s, and the fact that the former colonies of Europe had begun to shake off their imperial yoke. These liberated nations found the need to reassess their histories, hitherto drafted by the colonizers.

The 1950s and the early part of the 1960s were the period when a great deal of optimism resided in science, and its ability to serve as the panacea for human suffering. Within this mindset, a positive linkage of science, technology, and development was considered irrefutable. Despite the shifting ground of the science and culture debate, another version of the diffusionist model was proposed by the historian of science George Basalla (1967, 1968) to understand and model the expansion of European science in a paper entitled 'The Spread of Western Science'. The model proved to be resilient for a number of reasons. The fact that it was clothed in the apparel of physical theory lent it a degree of legitimacy. Further, it provided historians of science just entering the fray of the history of science in the non-West a grid to frame their material. Finally, its prescriptions dovetailed with the agenda of developmental agencies at the time, for the model was analogous with the then current understanding of the transfer of technology.

However, parallel models were elaborated in other disciplines in the post-World War II era. These included the convergence thesis of Bell and others, according to which all industrializing nations arrive more or less at the same end-point or look more or less alike, thereby prefiguring the sentiments of the end of ideology theorists (Elzinga 1980). Meanwhile, in economics Rostow's stage theory of development and later the Club of Rome thesis to the limits to growth were premised on the postulate that Western growth models drawing upon the research base of modern science and the infrastructure of technology were paradigmatic for future development.[1] Thus, Basalla's model was structurally cognate with contemporaneous theorizations of science, society, and the economy. Despite its resilience, the model, like all good models, was challenged a decade later. In this essay, we re-examine the model given the current appreciation of the history of science in the non-West.

'THE SPREAD OF WESTERN SCIENCE' REVISITED

Following the tremendous intellectual and cultural reorientation of the social studies of science, Basalla's paper of 1967 requires another look, for in its own time it was possibly the exemplar of the diffusionist history of modern science. Basalla's paper on the spread of Western science seeks to provide a model for the expansion of European sciences in two senses: the one geographical, and the other cultural and cognitive. In the first sense it seeks to provide a generalized description of the expanding domain of modern Western science that was concurrent with the expansion of European imperialism. As the dominion of European science expanded, the cognitive content and theoretical edifice of modern Western science also underwent significant modifications—it is the latter process as well that Basalla attempts to integrate into his model. The explicitly stated premises of his paper, it could be argued, are the following: (*a*) the original home of modern science was in the nations of Western Europe; and (*b*) these nations of Western Europe were the arena of a 'Scientific Revolution' that altered the epistemological, institutional, and social architecture of pre-modern science. In this regard, Basalla is situated within the Big Picture historiography of the post-World War II era, a historiography whose genealogy extended into the core of the Needhamian project (Bernard Cohen 1985; Floris Cohen 1994; Cunningham and Williams 1993); however, unlike Needham (1969) who sought to understand the emergence of modern science, Basalla's quest related to the diffusion of science from western Europe to other parts of the globe, including eastern Europe, North and South America, India, Australia, China, Japan, and Africa. Driven by the idea that science was a constituent element of the civilizing mission, Basalla failed to recognize the problematic status of the diffusion of so-called modern science even within Europe. Most recent researches testify that even if the Scientific Revolution did take place in Europe, this involved a process of progressive change (Shapin 1996). Consequently, there were substantial time lags marking the emergence of revolutionary scientific ideas in certain regions of Europe and their adoption in other regions. Consequently, arguing from symmetry, explanations of the spread of science in the non-West as the outcome of uneven development are to be invoked in explaining the spread of science even within Europe (Fuller 1997: 108) .

The three-stage model proposed by Basalla (Figure 8.1) was sufficiently attractive in its own time. However, in the light of the diversity of scholarship available today, we naturally respond to Basalla's model with scepticism for it is premised on a culturally transcendent model of modern scientific institutions and the stages of its institutionalization. This model, evolutionary in its tenor (Macleod 1987), sees the expansion of modern science through the establishment of

Figure 8.1: A Schema of Basalla's Model

Premises:

(1) Modern science emerged in the nations of Western Europe.

(2) A Scientific Revolution occurred in these nations in the 17th century.

(3) Modern science subsequently diffused to non-Europe.

Methodology:

(1) Identify the vectors of modern science.

(2) Identify the kind of knowledge disseminated by these vectors within the non-West.

(3) Identify the mechanisms sustaining scientific research activity outside Europe.

Outcome: Stages in the institutionalization of modern science outside Europe

Stage (1): Non-scientific society or nation provides source for European science.

Stage (2): Colonial science.

Stage (3): Completion of transplantation of modern science and commencement of struggle to achieve an independent scientific tradition.

modern scientific institutions. In order to map the diffusion of modern science outside Europe, Basalla finds it necessary to examine three dimensions of this process: (*a*) identify the vectors of modern science; (*b*) identify the 'kind of knowledge' and 'which knowledge' these vectors purveyed to non-Western regions; and (*c*) identify the mechanisms through which scientific research systems and research activity were sustained outside the ambit of Western Europe.

Basalla's model stimulated attempts to open up the Pandora's box of science and imperialism, but did so inadvertently. The polemics that followed its publication countered the model from two principal perspectives: those who assumed the 'science as cultural universal' framework proffered historical evidence that challenged the model on the grounds of empirical inadequacy or interpreted the same evidence differently within the same framework or proposed an interpretive variation of it. On the other hand those situated within the cultural studies of science argued for a move from the growth of science to the generation of knowledge, and those within the genre of transmission studies sought to interpose translation models into the history of science.[2]

THE EUROCENTRISM OF BASALLA'S THREE-STAGE MODEL

Amongst the many theories of transmission in the history of science there are those that are Eurocentric in scope. The latter theories, problematic though they are, find a place in academic institutions and often certify informed expert opinion. Any historiographic reflection must ask the question how unexamined and sometimes prejudiced historical claims have gained credence in European historical thought (Blaut 1993: 9). Furthermore, if our theories of history are value laden, and we recognize that they are, then it is relevant to ask whether these Eurocentric assumptions underlie Basalla's model.

The model itself is not evolutionary but quasi-evolutionary, since the trajectory leading to the emergence of modern science embedded within its related institutions and practices is a privileged one. For in the paper Basalla specifies the institutional interventions required to ensure the installation of such a tradition into non-Western societies, or even Western societies. In historiographic terms, this model is underpinned by an over-determinationist theory of history (Fuller 1993). Furthermore, the privilege accorded to modern science is premised on a theory of societal and scientific progress. In fact, at ever so many junctures in his argument, Basalla is vulnerable to the grand conflation (Gould 1989: 43), wherein the ideas of social progress and scientific progress are conflated with the idea of biological evolution. Contemporary researches on the social nature of science would suggest that Basalla's model is vulnerable to the charge of Eurocentrism on the one hand, and on the other of ignoring the process of cultural reception. Basalla's diffusionist model influenced early studies on science and imperialism, but as Petitjean (1992: 4) indicates, other problematics distinguished later studies, such as those relating to cultural reception, the conditions of the production of modern science, and aspects of the integration of modern science in non-Western cultures.

In fact, the influence of Basalla's model began to decline in the 1980s when doubts were cast on its adequacy. The first set of questions arose from the recognition of its Eurocentric definition of science that could not accommodate cross-cultural exchanges between 1450 and 1800. During the period under consideration 'more' non-European knowledge travelled to Europe than is often acknowledged. There is evidence to suggest the influence of Amerindian knowledge on European map making, and there are indications of Portuguese doctors learning from Indian *vaids* and *hakims* (Storey 1996: xviii). Second, critics suggested that Basalla was preoccupied with the spread of modern Western culture throughout the world without recognizing that the meaning of science changed across cultures, and within cultures across time (Storey 1996: xv). In his effort to recast the Basalla model, Macleod (1987: 227) faults it on six counts: (*a*) it was oblivious of the cultural environment of very diverse societies; (*b*) it totalized the scientific ideology of the West that was disseminated to the non-West; (*c*) it did not explain how political and cultural factors alter the shaded areas between the three phases; (*d*) it did not explain how science comes to occupy the centre stage of modern culture through its relation with technology and modern culture; (*e*) it did not account for the emergence of neo-colonialism; and (*f*) it did not account for interdependencies that have contributed to the plight of the third world. Using an economic metaphor, Sagasti elegantly summed up the model as proposing that the invisible hand of science would guide the responsible to independence (ibid.: 227). In metatheoretic terms, the model explained or accounted for so little, and yet its scientific aura was reflected in scholarly studies that invoked it in order to develop a cartography of the institutionalization of Western science.

Basalla proposed three institutional phases in the spread of Western science that in turn stimulated a great deal of discussion regarding the process of expansion of European science. This discussion and polemic related to the specificity of the experiences of non-Western societies, settler societies, and the nations of Eastern Europe. These discussions within the framework of Big Picture historiography touched on the following constitutive elements of the model:

- the extent to which the growth of scientific institutions and the diffusion of modern science in non-Western societies simulated the phases set out by the model; the other side of the issue being whether the model was relevant at all to the societies under discussion
- assuming that the model was relevant as a heuristic device, what was the differential overlap between the various phases set out by the model?
- as a generalization it was thoroughly Eurocentric, and was unable to distinguish between different colonial experiences

A specific examination of the stages of the model reveals that the distinction between Basalla's stage of colonial science and the next phase involving the formation of an independent scientific tradition is fuzzy. Basalla was aware of the problem, for he wrote: 'Colonial science contains, in an embryonic form, some of the essential features of the next stage' (Basalla 1967: 611–22). But suppose we were to argue, as Basalla acknowledges might be the case, that the third phase of an independent scientific tradition and the second phase of colonial science could coexist within the same temporal period. In this situation they could as well interlock with the interests of two very different constituencies: the former focusing on the establishment of a national scientific research system, and the latter on the economic and administrative interests of the colonizer. If this were the case, the colonial science research system would be relatively autonomous of the independent scientific tradition in the colony. The relationship between the two might be of the nature of lateral linkages that would ensure the flow of technical practices, skills, and information. While Basalla fittingly suggests that retrospective reconstructions of the independent scientific tradition overplay its realization and underemphasize its dependency on an older scientific tradition, he himself ignores the dialectic between different knowledge systems in non-Western societies (Rashed 1989: 205–7) for he visualizes the establishment of the modern scientific research system as totally displacing ancient and traditional knowledge systems, institutionally and cognitively.[3] An interesting study in its own right would be to examine the literary devices employed by Basalla in addressing the traditional knowledge systems of non-Western societies, for these do not appear to be issues relevant to the history of science, except as obscurantist traditions opposing the forward march of modern science.

ELEMENTS OF DIFFUSIONISM

In the three-stage model the first phase of transmission involved the presence of European scientists in the colonies, busy collecting data and conveying this data back to Europe. This science largely related to geographical exploration and natural history. Basalla pointed out that this phase was not restricted to an 'uncivilized country' that was the site of European settlement.[4] Explorers of this type whom Pyenson (1993a) refers to as vectors of cultural imperialism were socialized in the epistemic idiom of modern science. In specifying the distinction of this science from other knowledge forms and practices, Basalla drew upon characteristic 19th-century distinctions between Western modern science and Oriental science. In other words, he saw the collection of data as no longer organized around metaphysical or speculative principles as structured by the epistemic grid of modern science. Subsequently, Basalla discusses the transplantation of this science in the United States and Australia. This phase more easily fits into the transplantation of modern science in settler societies, wherein slash-and-burn colonization tended to preclude the possibility of dialogue with the Amerindians or the Aborigines of Australia.

On the other hand, when Basalla addresses the case of India and China his framework draws upon the debate between the ancients and moderns from the Enlightenment period. In this case the focus of analysis shifts towards how ancient societies responded to modern knowledge. But even here, Basalla's interest resided not in analysing cultural reception or cultural redefinition (Habib and Raina 1989), but in determining whether these cultures responded positively to the advent of modern science; and when they did not in ascertaining what sorts of beliefs and social organization posed impediments to the institutionalization of modern science within these societies. It was here that Basalla's inability to address the politics of knowledge becomes evident.

For Basalla in the phase of colonial science, scientific activity graduates to a higher theoretical level of activity. But this science is recognized as dependent science; and Basalla clarifies that colonial sciences is not a pejorative term and does not 'imply the existence of some sort of scientific imperialism whereby science in the non-European nation is suppressed or maintained in a servile state by an imperial power' (Basalla 1967: 611–22). It is the problematic nature of Basalla's depiction of this phase that compelled Macleod into not merely revising the three-stage model of Basalla in favour of a more federal one, but of pointing out the different meanings of imperialism and imperial science. Consequently, Basalla's becomes one of the possible frameworks for addressing the specificity of the expansion of European science. Imperial science, Macleod (1987: 220) reasons, so considered acquires different meanings when visualized from the centre and the periphery. The meanings associated with the term imperial science co-evolved with the project of imperialism itself and possibly reinforced each other.

Hence, in a number of cases, historical and political factors relating either to the assertion of colonial power or to the assertions of political sovereignty alter the sequence of shifted curves of the Basalla model. Instead, we have a set of highly clustered curves such that each of the phases ended up ensconced within the other. The sequence of shifted curves of the three-stage model is a representation of a linear discourse on the evolution of the institutions of modern science, ignorant of the political factors shaping the scientific research system, or the social organization and belief structures playing a role in its emergence. But suppose we were to accept that even within this model several different phases were to exist concurrently, in different states of evolution. Under the circumstances, Basalla posited that the spectrum of activity during this phase widely expanded, but given the dependent nature of colonial science the colonial scientist was unable to open up a new area. The colonial scientist could be of non-Western origin trained in Western science, a Western settler raised in a colony—the defining features included the institutional affiliations that extended beyond the territory where his scientific work was conducted (Basalla 1967: 8).

But this accords too much power and influence to the centre, or the metropolis, depending on whether we speak the language of centre–periphery or metropolis–province. For, historically, we do have a number of cases where the periphery has been the source of important new ideas in the theoretical sciences, examples being those of M.N. Saha and S.N. Bose from India, in theoretical astrophysics and quantum theory respectively. In Basalla's sociology of science, the likelihood of idea hybridizations at the periphery is discounted, whereas more recent researches suggest that the possibility of idea hybridizations at the periphery are greater than at the centre since the pressure of conformity within institutions, academies, and disciplines is lower (Chayut 1994: 28). There is, furthermore, substantial historical evidence of hybrid entities redirecting efforts of scientists at the centre. The point then is to ask why it is that we have so little evidence of the same. The question in which case will have to be rephrased: namely, even in instances of idea hybridizations at the periphery, the possibility of stabilizing the research network is low on account of the paucity of institutional, technical, and material resources (DeVorkin 1994: 158). This is not to say that Basalla wasn't aware of the disadvantages of undertaking scientific research in a former colony.

But Basalla also lived through the Kuhnian revolution in the social sciences, and the effect of Kuhn is evident in his article inasmuch as he envisions science as a kind of activity embedded in particular kinds of institutional practices.[5] Thus, while never suggesting the notion of the radical sociology of knowledge that modern science could be an ethno-science (Harding 1994), Basalla recognizes the precarious situation of the colonial scientist working in a field in which the core of activity is located at the metropolises of science.[6] Colonial scientists are thus handicapped inasmuch as they can never benefit from the invisible colleges 'in which the latest ideas and news of advancing frontiers of science are exchanged' (Basalla 1967: 611–22). And this disadvantage persists, according to him, into our own times.

These insights, however, were not integrated into his broader thesis. The term colonial alludes thus to too many distinct experiences. While on the one hand the colony is a territory settled by members of a mother country, there also existed colonies such as India, Cambodia, and Nigeria that were the non-settler outposts of the empire. This inappropriate categorization runs its course when he discusses Lomonosov and Russian colonial science in the 18th century and American science at the time of Benjamin Franklin (ibid., 1967: 9).[7] We are thus forced to conclude that the term colonial for him was no more than a euphemism for any 'other' of Western Europe that produced modern science.

The institution of an independent scientific tradition in the third stage requires surmounting traditional resistance to its establishment. According to Basalla, unlike the colonial scientist who can exist simultaneously in second stage and third the sustenance of an independent scientific establishment would require the elimination of any kind of opposition from traditional belief structures and interests. Basalla's sociology of science is founded on a faith in scientism. The opposition between traditionalism and the values of modern science are a core element of his sociology that prompts his remark concerning the institutionalization of science in China: 'Confucian ideals were decisively challenged and gradually replaced by value systems closer to the spirit of Western science' (ibid.: 611–22). But this stark dichotomy tends to make Basalla oblivious to the cultural hybridizations that illuminate the contemporary history of theoretical physicists, with the iconic presence of figures such as Sin-Itiro Tomonaga, Hidekei Yukawa in Japan, S.N. Bose or Srinivasa Ramanujan in India, or even Schrödinger, Pauli, and Mach in Vienna (Bitbol 1990; Kanigel 1991).

POLITICAL COLONIZATION OF SCIENCE

There are two further themes that arise in any contemporary reading of the Basalla paper. The first has to do with Basalla's inability to appreciate the distinction between the responses of China and Japan to modern science and modernization on one side, and the response of India on the other hand. The 'East' or the 'Ancient' world appears as an undifferentiated whole in his scheme. The diffusion of modern science in China and Japan was strongly determined by ruling-class interests at the time, and the predisposition of dominant elites to modernization and the advancement of the sciences. Much later the rivalry and competition between China and Japan had an equally important role to play. Furthermore, it has been argued that in the case of China and Japan it is possible to allude to a process of modernization in the previous centuries that was not aligned with the evolutionary trajectory of modern science and technology. Will (1995) refers to this process as 'modernisation less science'.[8]

Besides, political sovereignty or its absence was an important factor governing the absorption of technology. Inkster (1988: 405), in a comparative study on China, Japan, and India, identified three factors facilitating Japanese absorption of technology from abroad: (*a*) the labour-saving character of technology; (*b*) the utilization of existing social overhead capital; and (*c*) the encouragement of well-organized government. On the other hand in the case of India there were strong countervailing factors that worked against their absorption: (*a*) British interest in India was restricted to commerce, and the natural history enterprise; (*b*) consequently, the British did not establish institutions to produce skilled workers to man Indian or Western industrial enterprises; (*c*) imperial socio-economic policy did not compensate progressive indigenous groups for their loss of traditional instruction; and (*d*) the enclavist nature of the imperial economy was not conducive to the effective transfer of technology (ibid.: 417).

Hence, as recent scholarship on India has highlighted in the late 18th and 19th centuries, the loss of political and economic sovereignty produced a situation wherein the diffusion of modern science was more or less decided by imperial colonial policy (Baber 1996; Macleod 1987; Macleod and Dionne 1979; Macleod and Kumar 1995). Furthermore, the vectors of this knowledge censored even the scientific knowledge transmitted to India. Taking a point up from a more recent segment of the history of science, one can see that premier scientific societies such as the Royal Society of London functioned as the 'linch-pin of an empire wide system of patronage', bringing the science generated in the colonies under the supervision of metropolitan science (Home 1991: 151). While the Royal Society had a central place in the international patronage system of science, it nevertheless aided Britain in retaining a cultural hegemony in science over former colonial territories even after independence (ibid.: 152).

It was this interplay between knowledge and power that led even the historian of science S.N. Sen (1988) to ask why it was that in the 19th century, marked by the most rapid expansion of European science, the dissemination of the sciences in India was so tardy. Sen appropriately pointed out that the field sciences, including botany, geology, geomagnetism, geography, and astronomy, were transplanted in India in the 19th century, but the growth of physics, chemistry, and the basic sciences was incremental. The sciences that served the colonial need to survey and map the continent, and to ensure their governance and expropriation of resources from the colonies were rapidly instituted. It was only towards the last decades of the 19th century at the initiative of Indian scientists, both those trained abroad and those trained in India, that the demand for a scientific research system situated within the university became increasingly strident and political (Raina and Jain 1997). A bibliometric study on publications in physics from India between 1800 and 1950 revealed that the rate of growth of research publications in physics which were not promoted by the British was much more rapid than that of publications in the sub-disciplines of physics, in particular, geology and geomagnetism (Raina and Gupta 1998). Consequently, politically motivated interests of the developed world may pose impediments to what Basalla might consider the natural transmission of scientific knowledge and values. This inattentiveness to the politics of knowledge renders the model open to the criticism that it is inadequately instantiated, and explains how it draws settler, colonial, and Eastern European encounters with modern science under the same rubric. It is this decontextualization of science, sanitized from its political context, that accords science privilege, and when devoid of any sense of its practice renders it a kind of scientific imperialism (Nader 1996: 3).

But beyond the triumphalism of European science and the model's innocence of the cultural interaction between different kinds of knowledge, there are other aspects of the Basalla model that are found wanting from the perspective of the contemporary sociology of knowledge. The framework of the globalization of modern science (Schott 1993a, 1993b) is not sensitized in its present form to examine the knowledge of the disenfranchised and marginal social groups (Shapin 1994). Big Picture historiography suffers from this inherent optical scotoma: not only is Boyle's technician invisible, but so are knowledge forms that enrich the epistemic and cognitive dimensions of modern science but are marginalized politically and institutionally. Thus, Basalla's model, inasmuch as it carries the apolitical baggage of the Cold War era (Elzinga 1999; Fuller 1992), is devoid of the notion that modern science possessed an inherent cross-cultural comprehensibility (Blue 1999: 53) that was so essential to the Needhamian project from which it draws inspiration.

THE FUTURE OF DIFFUSIONIST HISTORY

In his conclusion Basalla concedes the mainly heuristic role of his model, which appears general enough, though at the time the paucity of comparative studies in the history and sociology of science could not have tested the generality of the model. The concluding section explicitly bears the philosophical underpinnings of the rest of the paper, inasmuch as it is founded on a strict internal–external dichotomy. Scientific and national styles are at best minor conceptual variations on the otherwise smooth evolutionary plane of scientific ideas. The 'social' only provides a backdrop for the efflorescence of scientific creativity, and in no way shapes the evolution of science. It is this founding mythology that has given universalistic credence to the Eurocentric history of science.

It could well be suggested that in addition to the studies on the transmission of Western science informed by the Basalla model, two other alternative explanatory frameworks have since been proposed. Studies informed by the Basalla model read science as a cultural universal, no matter what the geographic or cultural context within which it is transplanted, and imperialism is another universal altering the institutional destiny of science within these contexts. The focus of attention is how colonial scientists transplanted in the colonies established the suzerainty of Western science. Studies such as those of Pyenson pay attention to the ingredients of sciences themselves in detailing this history, while others highlight the administrative and political moves of colonial administrators, policy makers, and their native agents in furthering the agenda of science and imperialism.

The alternative approaches on the other hand analyse the process of 'cultural reception', addressing concerns such as the legitimation of science in different cultural spaces, the dialogue and confrontation between different knowledge systems, the politics of knowledge as introducing an element of epistemic drift (Elzinga 1993) in the practice of the sciences themselves, and most importantly, breaking away from any historiography founded on scientistic premises. A second stream amongst the alternative frameworks, which in a way shall more or less usher a major departure from conventional diffusionism, is founded on the cultural turn to science studies (Parusnikova 1992; Rouse 1987). This in turn throws up much deeper questions for the theory of social sciences and the project of history per se. For one, the history of transmission is much more comprehensive and replaces the traditional perspective of diffusion with that of translation. These approaches have challenged the Eurocentric history of science, and subsequent contributions might render Eurocentrism altogether irrelevant. For this to happen, as is becoming evident, the history of science will be transformed into the history of ever so many knowledge forms (Cunningham and Williams 1993). The anthropology of science has come to terms with treating science as any systematic way of generating knowledge, whether this covers the map-making procedures of the Amerindians or high-energy physics (Nader 1996). A dream uniting all those who believed in the emancipatory potential of science but subscribed to history of science premised on a positivist theory of science, was the relation of science to democracy.

While we have come a long way from the conventional philosophy of science, its dream of democracy and science may still be realizable.

However, in this proliferation of frameworks and interpretive possibilities we have to ask what happens to the crucial notion of the unity of sciences, when, as Hacking and Gallison have elegantly suggested, though independently, the cult of disunity prevails among historians and sociologists of science. But though this may be germane to the present essay as a larger philosophical problem, suffice it here to hastily suggest that 'something all too material resists' our constructions, or that constraints and obstacles limit what we might consider to be socially constructed.[9]

NOTES

1. This was pointed out to me by Merle Jacob of the Department of Theory of Science and Research, Göteborg University.
2. See previous chapter for a discussion on colonial science and translation models. In fact, Baber's (1996) is a recent book on the history of science in colonial India that departs from the Basalla model.
3. Rashed (1989: 207) points out that the act of translating knowledge from one system to another is structured by contemporaneous research trends, whose intention is never purely translational but of integrating one system with another. Such translations may also be motivated by the desire to surpass the knowledge that is being translated, and may in the process stimulate the extension of the translated knowledge.
4. Despite all his disclaimers to be politically correct, the term uncivilized was employed with reference to the American Indians and the aborigines of Australia, which reiterates a point made earlier in the essay.
5. Storey (1996: xv) suggests that Basalla had adopted the Kuhnian viewpoint that 'institutional pressures caused scientists to conform to dominant intellectual paradigms'. But neither Basalla nor Kuhn would have endorsed the view that externalities have an impact on the objectivity of scientific knowledge.
6. In an incisive paper Harding (1994: 302) argues that multiculturalism has posed three new questions relating to the natural sciences: (*a*) what is the extent to which modern science originated in Western culture? (*b*) are there culturally alternate sciences that are as universal as modern science? and (*c*) in what ways is modern science culturally European or Euro–American? The debate about ethno-science in a way arises from the recognition that science is both a way of categorizing the world as of defining itself to knowledge forms 'that are excluded' (Nader 1996: 3).
7. Thus, Basalla (1967: 6) clarifies: 'Phase 2 can occur where there is no

actual colonial relationship. This usage permits discussion of 'colonial science' in Russia or Japan as well as United States or India'.

8. The essays appearing in the volume of Hashimoto Keizô et al. make a well-argued case for re-examining the notion of modernization.

9. See the papers of Hacking, Gallison, and Pickering in Buchwald (1995).

9

Future Trajectories

The history of modern science dates back 300 or 400 years, depending on when historians or sociologically oriented historians seek to place the birth or origins of modern science (Crombie 1994; Cohen 1985; Shapin 1996), and historians of ideas seek to place the birth of modernity itself (Toulmin 1990). And yet the paradox that confronts the historiography of science is that 90 per cent of all science that historians of science make it their business to investigate has been produced in the last 50 years, while the majority of historians are devoted to studying the science produced in previous centuries (Söderqvist 1997). The present of science thus overwhelms the historian of science studying the past, and simultaneously de-skills the historian untrained in the sciences (ibid.: 9–10)[1]. The discipline of the history of science falls victim to the specialization of the sciences. Before proceeding to examine some of the questions and concerns of historians of science in the new millenium, it would be fitting to examine the history of the history of science as a discipline, and the changing nature and organization of science over the past three decades on which it would be possible to platform the future of the discipline.

The history of science as a modern discipline, given that its earlier variants date back to the ancient Greek and Arabic scholarly traditions, sought to chronicle the idea and development of the human mind, and Enlightenment intellectuals saw in the progress of the sciences, and mathematics in particular, an exemplar representative of human development (Crombie 1994). Attempts to transcribe this history, that they sought to give as universal a canvas as possible, brought them to consider the sciences of other cultures and civilizations as well (Peiffer, forthcoming). Thus, very early in the history of the disciplinary history of science a knowledge of the sciences of the non-West was constitutive of the discipline. Towards the second half, and in particular the last decades, of the 18th century, as the identity of modern nation states commenced stabilizing in Europe, the discourse on the past of the sciences was, among other factors, steered as much by the process of institutionalization of science, and the cognitive and institutional differentiation that characterized it (Laudan 1993: 1–34). In addition, there existed the need to circumscribe national identities themselves, and the priority dispute became its most significant marker. Science in these European nations came to be considered the degree of advancement of a nation (Adas 1990).

Overlapping these developments was an instrumental strain within the Enlightenment, drawing its intellectual capital from the evolution of mathematics in the 18th century. This vision found its most elaborate articulation in Condorcet, and that in the post-revolutionary context of the 19th century was rearticulated as a theory of social evolution by Comte (Liedman 1997). By the middle years of the 19th century two strands of the intellectual and cultural legacy of Europe from the previous centuries, steered by the nation state and new institutions of science, crystallized in the paradigmatic works of Whewell and Comte on the history of sciences. Within the frames of positivist and inductivist science that sought to chronicle the progress of the human mind, at best epitomized in the history of sciences, a strong Eurocentric formulation emerged. This formulation was subsequently institutionalized within pedagogy and university curricula, and acquired the dimensions of a mental-scape that appeared for long difficult to surpass (Blaut 1993).

In most of the history of science produced till the 1930s the question of primary importance was to understand the origins of modernity or that of the origins of modern science, which also meant that social historians were grappling with the issue of the conditions that shaped the emergence of the scientific revolution in 17th century Europe. These investigations were informed by a comparative perspective that simultaneously sought to understand the non-emergence of modern science in the non-West (Cohen 1994). The image of science as a cultural universal set the frame for examining the history of science in the non-West (Cunningham and Williams 1993: 407–32). This image came to be contested from a diversity of perspectives and has been adequately discussed in the literature. A subject of current interest is the transformation of science and the images of science that circulate within the community of meta-scientists. A brief discussion would enable a glimpse of the directions the historical study of the sciences in India is likely to take in the future.

The history of science is shaped in important ways by the dominant paradigms prevailing in the world of science. Hitherto, both the philosophy of science and the history of science took as their exemplar and subject of study the rapid growth of the science of physics and mathematics on which the former's growth was dependent (though the reverse was also the case). The Kuhnian turn to the history of science, or the social turn, as it is referred to, though Kuhn himself never conceived it that way, was an outcome of a deep engagement with the historiography of the scientific revolution inasmuch as it concerned physics and astronomy. In the present context much of the debate on the notion of techno-science reveals how contemporary ideas of Big Science shape our current understanding of science as well as the historical categories we deploy to look at the past. However, over the past two decades the physicalist paradigm has made way for the ascendancy of the information science and the biological sciences paradigms. This is consequent to rapid advances in the computer sciences and information sciences on the one hand, and the biological sciences on the other. This is likely to alter the focus of concerns of historians of science, now looking into more recent segments of the history of science, and may as well shape how they look at the history of the life sciences in the past.

In addition, the rumblings in the international sphere of political economy, especially related to intellectual property rights, is likely to trigger off a host of new priority disputes relating to biodiversity, agriculture, and the global commons. Multinational corporations have initiated the bio-prospecting of ethno-botanical knowledge in the third world, and through the intervention of experts this knowledge is incorporated into cycles of moneymaking in the North. Gradually, the third world is silenced into buying back these products after they have been repackaged 'on a Western dominated global pharmaceutical market' (Elzinga 1999: 73–113). The effects of changes within science and a co-produced global order and political economy that steer research programmes amongst disciplinary academic communities is reflected in concomitant changes within the scientific research system, both local and global. This prompts changes in the history of science or how historians are likely to conceive the subject of their study.

The emergence of a new mode of knowledge production is altering stable images (Gibbons et al. 1994) that have thus far underpinned the discourse on the history of science. Over the past few decades we have seen the gradual emergence of research institutes situated outside university contexts that have been traditionally the centre for knowledge production. In regions of the world where the university research systems were strong, this shift has not caused ample disturbance in the task of knowledge production. On the contrary, the new mode of knowledge production runs parallel to the university research system. Nevertheless, policy makers and technocrats have been promoting linkages between university and industry, in order to keep the former afloat in the light of cutbacks and the latter competitive. In other parts of the world this has eaten into the university research system. Even in countries like India, which normally take time to catch up with the rest of the world, it is becoming evident that most of the research activity of international calibre proceeds at a few research centres and institutions of national importance (Basu and Nagpaul 1998). Some of these are elite universities or are research centres that have been set up at some distance from traditional universities. What will this do to the history of science?

The history of science as it developed in the West from the renaissance onwards entered its subject matter with an ideal of scientific knowledge produced by individuals of rationality and genius situated in isolated towers of learning, such as the sites for the modern production of knowledge, namely, the university. Sociologists of knowledge have done much to establish a picture that looks upon the process of knowledge generation visualized as the collective production of knowledge.[2] This is evident in a departure from accounts fixated upon an epistemology that is individualist, and that on the contrary sees certified scientific knowledge as produced by knowledge generating communities informed more broadly by society, consequently social order and the order of the natural world are co-produced (Dennis 1997; Elzinga 1996). In addition, with the emergence of large teams and scientific and technological research networks, the image of the scientist working alone in his laboratory is gradually disappearing. For example, research papers that have come out of the European Centre for Nuclear Research (CERN) in Geneva have carried the names of about 300 authors. These changes will have an impact on historical narrative on two counts. On the one hand internal developments within social theory and external developments within the sciences will prompt a focus more on science as a social activity and a process of knowledge generation. However, heroic biography will continue to have its place, but one may even expect the intrusion of the social in a less trivial way. How? Popular accounts of science and attempts at popularization of science have often pleaded for the introduction of the human element into accounts of scientific discovery and especially scientific biography. But this has meant little more than the inclusion of some biographical details of the life of a scientist to make him look more human rather than illustrate how science becomes or is a social activity. In other words, the social process of knowledge production has never been addressed.

The discussion so far has related to the factors internal to the history of science as an academic discipline that will influence the character and themata of historical production. Some of these themes will come up for discussion ahead. Even the external factors discussed above are basically external within an internal account of the disciplinary history of science. Both these challenge the Humboldtian ideal of knowledge production that historians of science assumed as sacrosanct for almost 200 years. But the picture that is now emerging contends with the socially distributed nature of knowledge production and brings in a new set of actors into the discipline of the history of science (Douglas 1980: 80–83), which the generally conservative discipline of the history of science in India has been reluctant to admit.

The exception to this has of course been the history of the ecological and environmental sciences; those in any case are to be situated within a different epistemological programme of science, though practising ecologists appear to be divided between the natural history and thermodynamics paradigms. But the appearance of this other mode of knowledge production has brought in social movements, grassroots organizations working with artisans and rural technologies, computer hacks, and a range of actors situated at a diversity of institutional locations, not traditionally considered sites of knowledge production which have yet donned the role of knowledge generators. The history of science as a narrative of the production of knowledge of the natural world and how we act upon it and are in turn shaped by it, requires a revision in order to integrate the study of social movements and their impact of knowledge generation. This revision in the conception of science could compel historians of knowledge to examine other modalities of knowledge generation. In India at least this is already under way in the work of anthropologists, but the distance between the community of historians of science—still largely considered by the community of social scientists to be trapped within a positivist theory of science, and not without reason—and that of the social scientists is still far from being bridged.

The two external factors that have altered the trajectory of the history of science are those of post-colonialism and multiculturalism that in a significant way interlock each other. In fact, from a third-world point of view, it is now recognized that developments in post-colonial history, feminist studies, post-structural critical theory, and developments within the sociology of scientific knowledge have played a non-trivial role in furthering the possibility of global history (Frank 1998; Harding 1998). This possibility has arisen because of the epistemic convergence of the way science is conceived and reinterpreted. But here we have to bring in the social factors that prompted these developments in the social studies of science.[3]

I do not wish to get into the diverse meanings of post-colonialism, but in the Indian context refer to it as an address marking the era after the end of British colonial rule. It is during this period that the domain of history becomes an important field of contest, and historical attempts were initiated to understand the history of the nation anew. During this period the history of sciences was also given a stimulus in the country, partially in an attempt to legitimate the growth of scientific institutions and state funding in science after the passing of colonialism (see chapter 6). This history was stimulated on the one hand to contest the Eurocentric history of science, which naturally during the early years did acquire a nationalist tinge. But, more importantly, these efforts prompted studies from a diversity of perspectives to understand why the scientific revolution did not occur in India, or what the obstacles to the advancement of the sciences in India were during the period of early modernity. However, over the decades, from the perspective of cultural theory, the history of scientific institutions proper, the sociology of science, and economic history, greater attention came to be placed on the impact of colonialism on the knowledge systems of India, and the institutionalization of modern science in the country.[4]

The interrogations of the cultural theorists on the one hand and those coming in from the sociology of knowledge and the politics of science both in India and abroad began questioning some of the models of the transmission of scientific knowledge that had been the staple fare of an earlier generation of historians of science.[5] Furthermore, they also confronted at an epistemological level the received definition of science, and opened up the debate to a more diverse and broader notion of science. The essence of the discussion was the demarcation problem that was essentially seen as one of drawing boundaries between disciplines and excluding others. Epistemological questions were thus opened up within a debate on social theory. What for a long time had been a debate within Northern academe on national scientific styles was reincarnated in the post-colonial environment as a debate on alternative sciences and cultural assimilation of science, and was incarnated in the West in other conceptions of science within the West that had been marginalized (Easlea 1980; Merchant 1980).

Multiculturalism was an offshoot of post-colonialism and the phenomenon of failed states in the post-colonial world, witnessed socially in the migration of populations from the former colonies to the developed world. In a sense, multiculturalism as a pedagogic movement in the North embodies a modality of coping with the changing chromaticity of Western societies over the last three decades. Furthermore, in the United States where the multicultural debate in the realm of science education is the most vocal, it becomes evident that school curricula can no longer soft-pedal the old Eurocentric history of science for reasons relating to democracy.[6] Furthermore, educationalists have to contend with the different upbringings of students from a diversity of cultural backgrounds. And each of these cultural constituencies demands a place in the sun. This has thrown up new research concerns for the history of sciences— concerns that feed into the pedagogy of science education, but mediated through developments in the cognitive sciences and cognitive learning.

The widespread familiarity with democratic politics and democracy itself as an organizing principle has generated a new set of concerns for historians of science and has caused even the most staunch defendants of the idea of the privileged status of Western culture as well as those still committed to Eurocentric history to hedge their accounts a little. It is an irony that in the early decades of the 20th century political theorists such as Dewey and others sought to nourish political theory by the norms of democracy as encountered in the world of science (Fuller 1997). However, the manner in which universality was constructed and modernity privileged within the historical discourse on science partitioned the world into the modern and those who had to be civilized into the democratic theory and modernized. This idea has been challenged by historians themselves.

Needham, for one, had worked towards an ecumenical history of science that would recognize the contributions of different civiliza- tions and cultures to the growth of modern science. The picture itself was severely limited and, as Chemla (1999) has pointed out, the Needhamian picture excluded streams that did not join up with the river of modern science. But in the Needhamian project, which con- tinued to evolve till Needham's death, there was an ongoing attempt to construct a theory of science drawing upon democratic theory, whose elements Blue (1999) has constructed as the principle of epis- temological egalitarianism, according to which 'knowledge is in theory communicable across cultural borders and that persons of any cultural background are in principle capable of utilizing it'. Conse- quently, science tends to acquire the potential for global social inte- gration. The rewriting of the history of science prompted by the Needhamian commitment to epistemological egalitarianism has in- spired interesting research in the subject.

However, the opposition to Eurocentrism and the inability to effectively engage with Western hegemony in the domain of interna- tional affairs, or the poor performance of former colonial govern- ments back home, has provoked reverse commentaries that mirror the chauvinism of Eurocentrism in the history of science. A feature of this version of history is that it seeks to claim priority of discovery for every scientific theory or invention of merit either in the Arab- speaking world, India, or China. Despite these developments there are historians of science committed to understanding the process of evolution of scientific ideas rather than merely pinning down ques- tions of priority. But it is likely as neo-liberal regimes are imposed on developing countries, and as the pressures of globalization further exacerbate the politics of identity, that the parochial genre of the history of science will continue to prosper for some time to come.

Some of these strains are likely to draw mileage from the criticism of science that is currently fashionable in order to challenge the epistemic hegemony of modern science and in the process centre- stage some of their own claims. The picture is complicated by the politics of GATT and intellectual property rights as transnationals attempt to steer international law and governments to gain financial advantage for themselves while trampling upon the global com- mons. This requires that we nuance our appreciation both of the Needhamian ecumenical picture and multiculturalism. Chemla sug- gests that the Needhamian picture with its multiple origins of science is of crucial importance when challenging the claims that science is essentially European. But we cannot subscribe to this picture in a changed political economy that coerces nations into accepting that the benefits of science will be dependent upon past contributions. Against this backdrop, the idea that history can confer rights could legitimate the propagation of an inegalitarian order (Chemla 1999: 238).

Amongst the many challenges to the standard theories of science are those that take recourse to some version of cultural or judgmental relativism that are as problematic as the theories of science that they challenge. The crucial problem for social theory is to broad-base the notion of rationality and practices relating to different ways of acting on the natural world that do not fracture this discourse any further. Already within the sociology of sciences laboratory studies have motivated the deconstruction of science that has generated a prolif-eration of science into ever so many sciences, that scholars have begun to wonder if they are speaking of the same object called science (Gallison 1996). The problematic has been recognized by feminist scholars such as Haraway and Harding who have been arguing for an orthogonal vision and a strong objectivity that would in addition integrate some of the insights of post-colonial science studies into the history of science. The concept of strong objectivity proposed by Harding (1992: 82–101) requires identifying social as-sumptions that (*a*) enter scientific research and conceptualize hy-potheses formation; (*b*) are shared by observers designated legitimate who constitute a collective, and go on to structure institutions and collective schemes of disciplines; and (*c*) distinguish between values and interests that impede the production of less partial or distorted accounts of nature and social relations. These values include those of fairness, honesty, and democracy.

What we are witnessing today is a pressure on the history of science to perform different functions within society. History can no longer be viewed as a museum exhibiting the dead wood from the past. The history of science in the 18th and 19th centuries was a chronicle of the progress of the human mind. For a number of scientists, the history of science provided the opportunity for an active engagement with the present of science—and this is nowhere more luminously reflected than in the history of mathematics, when only a few years ago we saw a 300-year-old problem solved. The history of mathematics, in any case, as demonstrated by historians of mathematics, cannot be framed within the historiography of scien-tific revolutions, and does not exhibit the stages of the Kuhnian cycle.[7] In the past, of course, history as much as the history of science served to confer identities for communities and institutions. This role of history should, with the passage of time, move to the fringe as disciplinary identities stabilize and re-emerge as new disciplines surface. On the other hand, history as heroic biography of leading scientists has often provided the humanist garnishing for science, particularly when it is over-sold as a disembodied object, endowed with an epistemic engine that generates truths about nature. As the teleology of progress comes to play a weaker role in underpinning the history of science, the latter shall play a more active role in the pedagogy of science, and perhaps also throw up insights for practising scientists.

Important developments in the area of economic and trade history proper are likely to have a significant impact on central concerns of the history of sciences in the near future. The core problematic addressed by Weber was to understand the rise of capitalism in the West, which in another way was seeking a response to the question 'how the West grew rich'. This was logically related to the rise of modernity and theories of modernization. Modernization theory was premised on the scientific and technological revolution, and these concerns were related to the fundamental Needham question about the factors responsible for the rise of modern science in the West. Thus, we see a cluster of problematics that provided a thematic unity to the sociology and history of science from Weber to Needham.

In recent years the historiography of modernity has come in for severe questioning. Of the many objections posed, three are most important from our concerns relevant to historians of science working on the history of sciences of India. (I do not wish to use the term Indian historians of science, for the community of historians working on the history of sciences in India extends beyond the geopolitical boundaries of India). The first relates to the tenability of the relation of science to the project of modernization per se. For one, it is now felt that there was a period of humanism and openness that preceded the scientific revolution by at least a 100 years, and that the period of the scientific revolution was really one of greater close-mindedness than is normally imputed to the period (Toulmin 1990). Second, the historiography of sciences of East Asia has for long worked within a paradigm of 'modernization less science'.[8] In these historical studies there is a recognition of processes of modernization in East Asia, and China and Japan in particular, that were not pivoted on the scientific revolution. Some economists refer to this as economic modernization. Similarly, recent researches on the period of early modernization indicate that the global history of the past 400 years has been witness to 'the self-evident phenomena of the multiplicity of modernities'.[9] These developments force us to reconsider the singularity of the emergence of modernization in the West. Unlike in East Asia the historiography of science in India has still to put modernization under parentheses. Within the domain of history proper, historians have often questioned the historiography that sees the pre-colonial period as one of decline and decadence (Pannikar 1980). The world systems theorists—Braudel, Wallerstein, Frank, and others—have long been debating when the world economic system came into existence (Frank 1998). And while there continue to be a host of debates concerning the actual emergence of this system, one thing more or less appears to be settled, at least for the time being: that the system existed long before the onset of modernization or the scientific revolution. This pulls the rug under the Eurocentric history of science, and stimulates the emergence of globological approaches that are avowedly multicentric, though I personally believe that whether that makes them necessarily multicultural is still a matter of debate.

I would like to close this very brief review with a summing up of what has happened and what we as historians of science committed to some version of an ecumenical picture of the advance of science would have to face up to. In the recent past sociological approaches to the history of science have played an important role in revising our conception and popular images of science. Some of these reconstructions have been contested by scientists and have prompted what has come to be called the 'Science Wars'. The bugbear of the problem is not that sociologists see science in context and scientists should not be averse to notions of sociological relativism. The problematic concern here is that of epistemological or judgmental relativism (Harding 1992). And there are any number of sociologists of science who find this problematic. But apart from these wars over academic turf, new knowledge forms and actors hitherto excluded from the history of sciences proper, and here I am not referring only to the ethnosciences, have been admitted to the discipline of history. This is a positive development that any version of liberal historiography would find difficult to talk itself out of. On the other hand we are confronted with a bit of a paradox. Empirical and laboratory studies of science have so broadened the question of the method of science and the epistemology of science that we are forced to wonder if there is anything that makes these diverse activities undertaken in scientific institutions part of one and the same object called science.

The changing political economy of former colonies, as they succumb to the pressures of neo-liberalism and are drawn into the bandwagon of globalization, provokes a backlash in the form of the politics of identity and ethnic conflict. In each of these cases these newly resurgent groups seek to appropriate the public space seeking legitimization for their claims from an imagined history and rejecting prior historical reconstructions as being coloured by the colonizers' prejudices and imperial intentions. How long this tendency is likely to last depends upon how these societies respond to the pressing economic crises that afflict them as well as how they negotiate their way through globalization. In the long run, however, jingoism is not going to pay and will not hold the alleged constituencies of its proponents together. But before that happens, there is the likelihood that irreparable damage would have been inflicted on the social fabric, and some of the gains of liberal historiography would have been reversed. On the other hand, at the frontiers of scientific research, international collaboration, research networks, and research programmes now extend beyond national boundaries. The multinational character of scientific research programmes will enforce a revision of nationalist historiography, deflecting its focus to the generation of scientific knowledge. In any case, the tension between the globological and national accounts will persist for long. Thus, while the Big Picture of the history of science was problematic, it is time to repaint it in a new manner such that we keep a notion of situated universality, while recognizing the possibility that politics intrudes into the process of knowledge production.

NOTES

1. Söderqvist suggests that historians of science have failed to address the history of contemporary science since it requires a familiarity that is not possible without professional scientific training. And even if that obstacle could be overcome, then with familiarity of technical details arises the danger that the historian becomes partisan to the science he or she is writing about. The future historian of contemporary science will have to walk the tightrope of being 'scientist and historian in one person'.

2. Furthermore, recent debates in the sociology of scientific knowledge and sociologically informed history deal with how deeply contemporary researches into the nature of the world are mediated by devices as

well as mediations at a number of other levels. This has raised the question of the social nature of our constructions of reality, and prompted debates as to where the social stops and the non-social commences. The sociologist of science Karin Knorr Cetina explains: 'Scientists do not ... interact with the world directly; they interact with, for example, what other scientists have said about the world. The concepts in terms of which they think are taken out of the literature. The interpretations which they impose on their experimental results are interpretations that have been established by other scientists or by themselves, ... what is going on it is not the world which 'appears' there in any pure sense, but scientists interacting with each other, with the literature, with established knowledge, with what you could possibly claim to be based upon established knowledge, extending it' (Callebaut 1993).

3. This is not the place to go into post-structural critical theory and social constructivism, for that has been extensively discussed in the literature, see Harding (1998); for a discussion of the impact of developments in the social theory of science on history of science and vice versa, see Pestre (1995).

4. See Chapter 2.

5. See Chapter 8.

6. See issues of the journal *Science & Education: Contributions from History, Philosophy and Sociology of Science and Mathematics.*

7. See Richards (1995). For the point of view that mathematics is not shielded from revolutionary change, see Grabiner (1985). But even Grabiner admits that revolutionary change is mathematics is not as 'destructive' as in the other branches of science.

8. See Will (1995).

9. See Wittrock (1998). In the same issue see the papers by Shmuel N. Eisenstadt and Wolfgang Schluchter on the comparative view of modernities; Sheldon Pollock on the vernacular millenium; and Sanjay Subrahmanyam on early modernity in South Asia.

Bibliography

Abrol, Dinesh. 1995. 'Colonized minds or nationalist scientists: The 'Science and culture group', in R. Macleod and D. Kumar (eds), *Technology and the Raj: Technical Transfers to India 1700–1947*. London: Sage Publications, pp. 112–33.

Adam, William. 1865. *Adam's Report on Vernacular Education in Bengal and Behar, Submitted to Government in 1835, 1836, and 1838 with a Brief View of its Past and Present Conditions*. Calcutta: Home Secretariat Press.

Adas, Michael. 1990. *Machines as the Measure of Men: Science, Technology, and Ideologies of Western Dominance*. New Delhi: Oxford University Press.

Allport, Phil. 1991. 'Still Searching for the Holy Grail', *New Scientist*, 5 October, pp. 55–56.

An Approach to the Science and Technology Plan, National Council for Science and Technology, Department of Science and Technology, New Delhi, January 1973.

Anderson, Robert S. 1999a. 'Peter Blackett in India: Military Consultant and Scientific Intervenor, 1947–1972—Part One', *Notes Rec. R. Soc. Lond.*, 53 (2), pp. 253–73.

—— 1999b. 'Peter Blackett in India: Military Consultant and Scientific Intervenor, 1947–1972—Part Two', *Notes Rec. R. Soc. Lond.*, 53 (2), pp. 345–59.

Anderson, Walter Truett. 1994. 'The Moving Boundary: Art, Science and the Construction of Reality', *World Futures*, 40 (1–3), pp. 27–34.

Alvares, Claude. 1979. *Homo Faber: Technology and Culture in India, China and the West: 1500 to the Present Day*. The Hague: Martin Nijhof.

Arunachalam, S. 1979a. 'Why is Indian Science Mediocre?' *Science Today*, February, p. 8.

—— 1979b. 'Scientific Journals in India: Their Relevance to International Science', *Science Today*, March, pp. 45–50.

Baber, Zaheer. 1996. *The Science of Empire: Scientific Knowledge, Civilization and Colonial Rule in India*. New York: State University Press.

Bachelard, Gaston. 1971. *Epistemologie*. Paris: Presses Universitaires.

Bannerjea, D. 1990. 'Contributions of Sir P.C. Ray to National Development', *Journal of the Indian Chemical Society*, 67 (April), pp. 269–85.

Barthes, Roland.1982. 'The last happy writer', in Susan Sontag (ed.), *Barthes: Selected Writings*. London: Fontana/Collins.

Barnes, Barry, and Edge, David (eds). 1982. *Science in Context: Readings in the Sociology of Science*. Milton Keynes, Buckinghamshire: Open University Press.

Basalla, G. 1967. 'The Spread of Western Science', *Science*, CLVI, (5 May), pp. 611–22.

Basalla, George (ed.). 1968. *The Rise of Modern Science: Internal or External Factors*. Lexington, Massachusetts: D.C. Heath.

Basu, A. and Nagpaul, P.S. 1998. 'National Mapping of Science: A Bibliometric Assessment of India's Scientific Publications based on Citation Index (1990 and 1994)'. New Delhi: NISTADS Report Rep 248/98.

Bayly, C.A. 1996. 'Who Discovered the Cause of Malaria', *Times Literary Supplement*, October 11, p. 33.

Ben-David, Joseph. 1984. *The Scientist's Role in Society: A Comparative Study*. Chicago University Press.

Bensaude-Vincent, Bernadette. 1986. 'Mendeleev's Periodic-System of Chemical Elements', *British Journal of History of Science* 19, pp. 13–17.

Besson, Jean. 1992. 'Table Ronde', in Jean Dhombres and Bernard Javault, (eds), *Actes de Colloque 'Marcelin Berthelot: Une Vie, Une Epoque, Un Mythe*. Paris: pp. 141–43.

Bernal, J.D. 1939. *The Social Function of Science*. London: Macmillan.

—— 1954. *Science in History*. London: C.A. Watts & Co..

Bernal, Martin. 1987. *Black Athena: The Afroasiatic Roots of Classical Civilization*. London: Free Association Books.

Bernier, F. 1989. *Travels in the Mughal Empire: AD 1656–1668*. Delhi: Low Price Publications.

—— 1992. *Abrégé de la Philosophie de Gassendi*, 7 vols, Paris: Fayard.

Berthelot, Marcelin. 1885. *Les Origines de l'Alchimie*. Paris: Georges Steinhel.

—— 1898. 'Sur l'Alchimie Indienne', *Journal des Savants*, Avril, pp. 227–36.

Biagioli, Mario. 1996. 'From Relativism to Contingentism', in Peter Galison, and David J. Stump (eds), *The Disunity of Science: Boundaries, Contexts, and Power*. Stanford: Stanford University Press, pp. 190–206.

Bitbol, Michel. 1990. 'Schrödinger: Philosophe chez les Physiciens', *La Recherche*, Novembre.

—— 1996. *Schrödinger's Philosophy of Quantum Mechanics*. Dordrecht: Kluwer.

Blaut, J.M. 1993.*The Colonizer's Model of the World: Geographical Diffusionism and Eurocentric History*. New York/London: The Guilford Press.

Blue, Gregory. 1999. 'Science(s), Civilization(s), Historie(s): A Continuing Dialogue with Joseph Needham', in S. Irfan Habib and Dhruv Raina (eds), *Situating the History of Sciences: Dialogues with Joseph Needham*. New Delhi: Oxford University Press, pp. 29–72.

Bose, D.M. 1956. 'History of Science in India', *Science and Culture*, 21(8), pp. 395–402.

—— 1963a. 'History of Science in India: How it Should be Written?', *Science and Culture*, 29(4), pp. 163–6.

—— 1963b. 'Asiatic Society's Contribution to Science Studies in India', *Science and Culture*, 29(5), pp. 219–24.

Bose, D.M, Subbarayappa, B.V., and Sen, S.N. 1970. *A Concise History of Science in India*. Delhi, INSA.

Brush, Stephen G. 1995. 'Scientists as Historians', *Osiris*, 10, pp. 215–31.

Buchwald, Jed Z. 1995. *Scientific Practice: Theories and Stories of Doing Physics*. Chicago and London: University of Chicago Press.

Callebaut, Werner. 1993. *Taking the Naturalistic Turn or How Real Philosophy of Science is Done*. Chicago: University of Chicago Press.

Canguilhem, Georges. 1988. *Ideology and Rationality in the History of the Life Sciences* (translated by Arthur Goldhammer). Cambridge, Massachusetts: MIT Press.

Chandra, Bipan. 1966/1977. *The Rise and Growth of Economic Nationalism in India: Economic Policies of Indian National Leadership 1880–1905*. New Delhi: People's Publishing House.

—— 1980. 'Karl Marx: His Theories of Asian Societies and Colonial Rule', in M. O'Callaghan (ed.), *Sociological Theories: Race and Colonialism*. UNESCO.

Chandrasekhar, S. 1987. 'On Ramanujan', in George E. Andrews et al. (eds), *Ramanujan Revisited* (Proceedings of the Centenary Conference, University of Illinois, Urbana-Champaigne). San Diego: Academic Press.

—— 1991. *Truth and Beauty: Aesthetics and Motivations in Science*. New Delhi: Penguin Books India.

Chatterjee, Santimay. 1986. 'Acharya Prafulla Chandra Ray: The Growth and Decline of a Legend', in Santimay Chatterjee and Amitabha Sen (eds), *Acharya Prafulla Chandra Ray: Some Aspects of His Life and Work*. Calcutta: Indian Science News Association, pp. 1–30.

Chattopadhyaya, Debiprasad. 1959. *Lokayata: A Study in Ancient Indian Materialism*. New Delhi: PPH.

—— 1976. *What is Living and What is Dead in Indian Philosophy*. New Delhi: PPH.

—— 1978. *History and Society: Essays in Honour of Prof. Niharanjan Ray*. Calcutta: K. P. Bagchi & Company.

—— 1979. *Science and Society in Ancient India*. Calcutta: K.P. Bagchi and Company.

—— 1982. *History of Science in India* (2 vols). New Delhi: Editorial Enterprises.

—— 1986. *History of Science and Technology in Ancient India: The Beginnings*. Calcutta: Firma KLM.

Chayut, Michael. 1994. 'The Hybridisation of Scientific Roles and Ideas in the Context of Centres and Peripheries', *Minerva*, XXXII, 3, pp. 297–308.

Chemla, Karine. 1999. 'The Rivers and the Sea: Analysing Needham's Metaphor for the World History of Science', in S. Irfan Habib and Dhruv Raina (eds), *Situating the History of Sciences: Dialogues with Joseph Needham*. New Delhi: Oxford University Press, pp. 220–44.

Clark, W.E. 1937. 'Science in India', in G. T. Garratt (ed.), *The Legacy of India*. London: Clarendon Press, pp. 335–65.

Cohen, I. Bernard. 1985. *Revolution in Science*. Cambridge, Massachusetts: Belknap Press of Harvard University.

Cohen, H. Floris. 1994. *The Scientific Revolution: A Historiographical Inquiry*. Chicago: University of Chicago Press.

Colebroke, H.T. 1873. *Miscellaneous Essays, with the Life of the Author by his son, Sir T.E. Colebroke* (3 vols), London: Trubner & Co.

Collins, Harry and Pinch, Trevor. 1993. *The Golem: What Everyone should Know about Science*. Cambridge: Cambridge University Press.

Coomaraswamy, Ananda K. 1904. 'Report on Thorianite and Thorite', in Wyndham R. Dunstan (ed.), *Report on the Occurrence of Thorium bearing Minerals in Ceylon*. Colombo.

—— 1918. *The Dance of Shiva*. New York: The Sunwise Press (republished by Sagar Publications, New Delhi, 1971).

—— 1927. *History of Indian and Indonesian Art*. New York: E. Weyth.

—— 1943/1947. 'Eastern Wisdom and Western Knowledge', *Isis* XXIV. (This article was published in his collection of essays *Am I My Brother's Keeper*. New York: Asia Press, pp. 46–55. The page numbers in this essay refer to the latter version).

—— 1947. 'East and West', in *Am I My Brother's Keeper*. New York: Asia Press, pp. 66–77.

—— 1944/1947. 'Gradation and Evolution I', *Isis* XXV; (1947) 'Gradation and Evolution II', *Isis* XXXVIII.

—— 1989. *Time and Eternity*. Bangalore: Select Books (first published in 1947).

Cooter, Roger and Pumfrey, Stephen. 1994. 'Separate Spheres and Public Places: Reflections on the History of Science Popularization and Science in Popular Culture', *History of Science* XXXII, pp. 237–64.

Cozzens, Susan E. and Gieryn, Thomas F. 1990. 'Introduction: Putting Science Back in Society', in S.E. Cozzens and T.F. Gieryn (eds), *Theories of Science and Society*. Bloomington: Indian University Press.

Crombie, A.C. 1994. *Styles of Scientific Thinking in the European Tradition : The History of Argument and Explanation especially in the Mathematical and Biomedical Sciences and Arts* (vols I and III). London: Duckworth.

Crozet, Pascal. 1999. 'Modernisation of Science and Its History Outside Europe: Egyptian Projects in the Nineteenth Century', in S. Irfan Habib and Dhruv Raina (eds), *Situating the History of Science: Dialogue with Joseph Needham*. New Delhi: Oxford University Press, pp. 260–78.

Cunningham, Andrew and Williams, Perry. 1993. 'Decentring the "Big

Picture": The Origins of Modern Science and the Modern origins of Science', *British Journal of History of Science* 26, pp. 407–85.

Dani, S.G. 1993. 'Vedic Mathematics: Myth and Reality', *Economic and Political Weekly*, 31 July, pp. 1577–80.

Dasgupta, Subrata. 1999. *Jagadis Chandra Bose and the Indian Response to Western Science*. New Delhi: Oxford University Press.

Dasgupta, Surendranth. 1980. *A History of Indian Philosophy* (5 vols). Delhi: Motilal Banarsidas.

Datta, B.B. 1929. 'The Hindu Solution of the General Pellian Equation', *Calcutta Mathematical Society Bulletin* 19, pp. 87–94.

—— 1931. 'The Origin of Hindu Indeterminate Analysis', *Archeion* 13, pp. 401–7.

Dennis, Michael Aaron. 1997. 'Historiography of Science: An American Perspective', in John Krige and Dominique Pestre (eds), *Science in the Twentieth Century*. Amsterdam: Harwood Academic Publishers, pp. 1–26.

DeVorkin, David H. 1989. 'Henry Norris Russel', *Scientific American*, May.

—— 1994. 'Quantum Physics and the Stars (IV): Meghnad Saha's Fate', *Journal for the History of Astronomy* XXV, pp. 155–88.

DeVorkin, David H. and Kenat, Ralph. 1983a. 'Quantum Physics and the Stars (I): the Establishment of the Temperature Scale', *Journal for the History of Astronomy* XIV, pp. 102–32.

—— 1983b. 'Quantum Physics and the Stars (II): Henry Norris Russel and the Abundance of the Elements in the Atmospheres of the Sun and Stars', *Journal for the History of Astronomy* XIV, pp. 180–22.

Dharampal. 1971. *Indian Science and Technology in the Eighteenth Century: From Contemporary European Accounts*. New Delhi: Impex Publishers.

Douglas, Susan J. 1980. *Isis* 81, pp. 80–3.

Dubos, René. 1950. *Louis Pasteur: Free Lance of Science*. New York: De Capo Paperback.

Edwards, Paul. 1994. 'Hyper-text and Hypertension: Post-structural Critical Theory', *Social Studies of Science* 24, 229–78.

Elzinga, Aant. 1980. 'Models in the Theory of Science: A Critique of the Convergence Thesis', in Erik Baark, Aant Elzinga, and Bengt-Erik Borgström (eds), *Technological Change and Cultural Impact in Asia and Europe: A Critical Review of the Western Theoretical Heritage*. Stockholm, pp. 37–70.

—— 1984. Essays on scientism, romanticism and social realist images of science. Institutionen for Vetenskapsteori, Goteborgs Universtet, Report No. 143.

—— 1988. 'Bernalism, Comintern and the Science of Science: Critical Science Movements Then and Now', in J. Annerstedt and A. Jamison (eds), *From Research Policy to Social Intelligence*. London/New York: Macmillan Press, pp. 87–113.

Elzinga, Aant. 1993. 'Science as Continuation of Politics by Other Means', in Thomas Brante, Steve Fuller, and William Lynch (eds), *Controversial Science: From Context to Contention.* New York: State University Press.

—— 1996. 'The Historical Transformation of Science with Special Reference to Epistemic Drift', in Christoph Hubig (ed.), *Cognitio humana-Dynamik des Wissens und der Werte*, Akademie Verlag. pp. 529–56.

—— 1999. 'Revisiting the "Needham Paradox"', in S. Irfan Habib and Dhruv Raina, *Situating the History of Sciences: Dialogues with Joseph Needham.* New Delhi: Oxford University Press, pp. 73–113.

Elzinga, A. and Jamison, A. 1981. 'Cultural Components in the Scientific Attitude to Nature: Eastern and Western Modes', *Technology and Culture,* Occasional Report Series, No. 2, Lund University.

—— 1986. 'The Other Side of the Coin: The Cultural Critique of Technology in India and Japan', in E. Baark and A. Jamison (eds), *Technological Developments in China, India and Japan.* London: Macmillan, pp. 205–51.

Eslea, Brian. 1980. *Witch-Hunting, Magic and the New Philosophy: An Introduction to the Debates of the Scientific Revolution 1450–1750.* Brighton: Harvester Press.

Farrington, Benjamin. 1953. *Greek Science.* Middlesex: Penguin.

Feyerabend, P.K. 1987. *Farewell to Reason.* London: Verso.

—— 1991. *Three Dialogues on Knowledge.* Oxford: Basil Blackwell.

—— 1994. 'Art as a Product of Nature as a Work of Art', *World Futures* 40(1–3), pp. 87–100.

Filliozat, J. 1964. *The Classical Doctrine of Indian Medicine: Its Origins and Greek Parallels* (translated by D.R. Chanana). Delhi: Munshiram Manoharlal (published in French in 1949).

—— 1974. 'La Naissance et l'Essor de l'Indianisme', in J. Filliozat, *Laghu-Prabandah: Choix d'Articles d'indologie.* Lieden: E.J. Brull, pp. 1–32.

Forbes, Geraldine H. 1975. *Positivism in Bengal: A Case Study in the Transmission and Assimilation of an Ideology.* Calcutta: Minerva Association Publishers.

Fox-Genovese, Elizabeth and Lasch-Quinn, Elisabeth. 1999. 'Introduction', in Elizabeth Fox-Genovese and Elizabeth Lasch-Quinn (eds), *Reconstructing History: The Emergence of a New Historical Society.* New York and London: Routledge, pp. xiii–xxii.

Frank, Andre Gunder. 1998. *ReOrient: Global Economy in the Asian Age.* New Delhi: Vistaar Publications.

Fuller, Steve. 1992. 'Being there with Thomas Kuhn: A Parable for Post-modern Times', *History and Theory*, October, pp. 241–75.

—— 1993. *Philosophy, Rhetoric and the End of Knowledge: The Coming of Science & Technology Studies.* Madison: University of Wisconsin Press.

—— 1994. 'Toward a Philosophy of Science Accounting: A Critical Rendering of Instrumental Rationality', *Science in Context* 7(3), pp. 591–621.

—— 1995. 'On the Motives for the New Sociology of Science', *History of the Human Sciences* 8(2), pp. 117–24.

—— 1997. *Science: Concepts in the Social Sciences*. Buckingham: Open University Press.

—— 1999. 'Towards a Prologemena for a Global History of Science', in S. Irfan Habib and Dhruv Raina (eds), *Situating the History of Science: Dialogues with Joseph Needham*. New Delhi: Oxford University Press, pp. 114–51.

Gallison, Peter. 1996. 'Introduction: The Context of Disunity', in Peter Galison and David J. Stump (eds), *The Disunity of Science: Boundaries, Contexts and Power*. Stanford: Stanford University Press.

Gascoigne, John. 1996. 'The Study of Nature', in Knud Haakonssen (ed.), *The Cambridge History of Eighteenth Century Philosophy*. Cambridge: Cambridge University Press.

Gibbons, Michael, Limoges, Camille, Nowotony, Helga, Schwartzman, Simon, Scott, Peter and Trow, Martin. 1994. *The New Production of Knowledge: The Dynamics of Science and Research in Contemporary Societies*. London: Sage Publications.

Gillispie, Charles C. 1980. *Science and Polity in France at the End of the Old Regime*. New Jersey: Princeton University Press.

Giri, A. 1992. 'The Portrait of a Discursive Formation: Science as Cultural Criticism and Social Activism in Contemporary India'. NISTADS Report No. WP-69/92.

Gizycki, Rainald von. 1973. 'Centre and Periphery in the International Scientific Community', *Minerva*, XI(4), pp. 474–94.

Goldstein, Bernard R. 1996. 'Astronomy and Astrology in the Works of Abraham Ibn Ezra', *Arab Sciences and Philosophy* 6, pp. 9-21.

Gould, Stephen Jay. 1989. *Wonderful Life: The Burgess Shale and the Nature of History*. New York: W.W. Norton.

Grabiner, Judit V. 1985. 'Is Mathematical Truth Time-dependent', in Thomas Tymoczko (ed.), *New Directions in the Philosophy of Mathematics*. Boston: Birkhauser, pp. 201–14.

Graham, Loren. 1985. 'The Socio-political Roots of Boris Hessen: Soviet Marxism and the History of Science', *Social Studies of Science* 15, pp. 705–22.

Guay, Y. 1986. 'Emergence of Basic Research on the Periphery: Organic Chemistry in India, 1907–1926', *Scientometrics* 10(1–2), pp. 77–94.

Guillemain, Bernard. 1992. 'Marcelin Berthelot et le Positivisme', in Jean Dhombres and Bernard Javault (eds), *Actes de Colloque Marcelin Berthelot: Une Vie, Une Epoque, Une Mythe*. Paris: CNRS, pp. 109–11.

Habib, Irfan. 1971. 'Potentialities of Capitalist Development in the Economy of Mughal India', *Enquiry*, New Series III, No. 3, pp. 1–56.

—— 2000. 'History and Interpretation: Communalism and problems of

Historiography in India'. http://members.nbci.com/_XMCM/indo-window/ godown/history/Ihhistint.htm

Habib, S. Irfan and Raina, Dhruv. 1989. 'Copernicus, Columbus, Colonialism and the Role of Science in Nineteenth Century India', *Social Scientist*, 190–1, pp. 51–66.

—— 1992. 'The Discourse on Scientific Rationality: A Study of Master Ramchandra', in T. Niranjana, P. Sudhir, and V. Dhareshwar (eds), *Interrogating Modernity: Culture and Colonialism in India*. Calcutta: Seagull Books, pp. 348–68.

Hacking, Ian. 1996. 'The Disunities of the Sciences', in Peter Galison and David J. Stump (eds), *The Disunity of Science: Boundaries, Contexts and Power*. Stanford: Stanford University Press.

Hagendijk, Rob. 1990. 'Structuration Theory, Constructivism and Scientific Change', in S.E. Cozzens and T.F. Gieryn (eds), *Theories of Science and Society*. Bloomington: Indian University Press, pp. 43–66.

Harding, Sandra. 1992. 'After the Neutrality Ideal: Science, Politics, and Strong Objectivity', in Margaret C. Jacob (ed.), *The Politics of Western Science: 1640–1990*. New Jersey: Humanities Press, pp. 82–101.

—— 1994. 'Is Science Multicultural? Challenges, Resources, Opportunities and Uncertainties', *Configurations* 2(2), pp. 301–52.

—— 1998. *Is Science Multicultural? Postcolonialisms, Feminisms, and Epistemologies*. Bloomington and Indianapolis: Indiana University Press.

Hardy, G.H. 1992. *A Mathematician's Apology*. Cambridge: Cambridge University Press (first published in 1940).

Hasse, Raimund, Kruecken, Georg, and Weingart, Peter. 1995. 'Social Expectations and Internal Dynamics of Science: A Neoinstitutional Approach'. Paper presented at the *EASST/ERASMUS Workshop on 'Social Theory and Social Studies of Science'*, University of Bielefeld, 9–12 May.

Hayles, Katherine. 1989. 'Reconfiguring Literature and Science: From Supplementarity to Complementarity'. *International Conference on the History and Philosophy of Science*, Munich.

Headrick, D.R. 1981. *The Tools of Empire: Technology and European Imperialism in the Nineteenth Century*. New York: Oxford University Press.

—— 1988. *The Tentacles of Progress: Technology Transfer in the Age of Imperialism*. New York: Oxford University Press.

Hess, David J. 1997. *Science Studies: An Advanced Introduction*. New York and London: New York Press.

Hiebert, Erwin N. 1982. 'Developments in Physical Chemistry at the Turn of the Century', in C.O. Bernhard, E. Crawford, and P. Sorborn (eds), *Science, Technology and Society in the Time of Alfred Nobe*. Oxford: Pergamon.

Hobsbawm, Eric. J. 1973. *The Age of Revolutions: 1789–1848*. London: Cardinal.

—— 1988. *The Age of Capital: 1848–1875*. London: Cardinal.

—— 1993. *The Age of Extremes: The Short Twentieth Century: 1914–1991.* London: Michael Joseph.

Hodgkin, Luke. 1986. 'Mathematics as Ideology and Politics', in Lev Levidow (ed.), *Radical Science Essays.* London: Free Association Press, pp. 173–97.

Holton, Gerald. 1993. 'Can Science be at the Centre of Modern Culture', *Public Understanding of Science* 2, pp. 291–305.

Home, R.W. 1991. 'A World-Wide Scientific Network and Patronage System: Australian and Other Colonial Fellows of the Royal Society of London', in R.W. Home and S.G. Kohlstedt (eds), *International Science and National Scientific Identity.* Dordrecht: Kluwer.

Huff, Toby E. 1993. *The Rise of Early Modern Science: Islam, China and the West.* Cambridge: Cambridge University Press.

Inden, Ronald. 1990. *Imagining India.* Oxford: Basil Blackwell.

Inkster, Ian. 1975. 'Science and the Mechanic Institutes, 1820–1850: The Case of Sheffield', *Annals of Science* XXXII (5), pp. 453–65.

—— 1988. 'Prometheus Bound: Technology and Industrialization in Japan, China and India prior to 1914: A Political Economy Approach', *Annals of Science* 45, pp. 399–426.

Jain, A., Mohanti, S., Krishna, V. V., Haribabu, E., Jairath, V. K., and Basu, A. 1992. 'A. Scientific Communities and Brain Drain: A Sociological Study'. New Delhi: *NISTADS Report.*

Jamison, Andrew. 1982. *National Components of Scientific Knowledge: A Contribution to the Social Theory of Science.* Lund: Research Policy Institute, University of Lund.

Kaneko, T. 1987. 'Einstein's impact on Japanese intellectuals', in F. Glick (ed.), *The Comparative Reception of Relativity.* Dordrecht: Kluwer, pp. 351–79.

Kanigel, Robert. 1991. *The Man Who Knew Infinity: A Life of the Genius Ramanujan.* New York: Charles Scribner's.

Kaviraj, Sudipta. 1988a. 'Imaginary History'. Occasional Paper on History and Society Issues, Second Series, No. vii, NMML, New Delhi.

—— 1988b. 'Humour and the Prison of Reality: Kamalakanta and the Secret Autobiography of Bankimchandra Chattopadhyaya'. Occasional Papers on History and Society, IV, NMML, New Delhi.

Knight, David. 1992. *Ideas in Chemistry: A History of the Science.* New Jersey: Rutgers University Press.

Kopf, D. 1969. *British Orientalism and the Bengal Renaissance,* Berkeley: University of California Press.

Kosambi, D.D. 1985. *The Culture and Civilization of Ancient India.* New Delhi: Vani Educational Books.

Kripke, S. 1982. *Wittgenstein on Rules and Private Language: An Elementary Exposition.* Cambridge, Massachusetts: Harvard University Press.

Kumar, Deepak. 1995. *Science and the Raj.* New Delhi: Oxford University Press.

Lardinois, Roland. 1994. 'The Field of Indological Studies in France in and Around the Second World War'. 13th European Conference of Modern South Asian Studies, Toulouse.

—— 1997. 'Louis Dumont et la science indigéne', *Actes* 106–7, pp. 11–26.

Latour, Bruño. 1988/1993. *The Pasteurization of France*. Cambridge, Massachusetts: Harvard University Press.

Laudan, Rachel. 1993. 'Histories of the Sciences and their Uses: A Review to 1913', *History of Science* XXX(1), pp. 1–34.

Lelyveld, David. 1993. 'Colonial Knowledge and the Fate of Hindustani', *Comparative Study of Society and History* 35(4), pp. 665–82.

Leslie, Charles (ed.). 1976. *Asian Medical Systems: A Comparative Study*. Berkeley: University of California Press.

Lettres Edifiantes et Curieuses: Memoires de l'Inde. 1810. Tomes 11–15. Toulouse: Noel-etienne SENS, Imprimeur Libraire.

Liedman, Sven-Eric. 1997. 'The Crucial Role of Ethics in Different Types of Enlightenment (Condorcet and Kant)', in Sven-Eric Liedman (ed.), *The Postmodernist Critique of the Project of Enlightenment*. Poznan Studies 58, Amsterdam: Rodopi.

Lipsey, Roger. 1977. *Coomaraswamy III: His Life and Work*. (Bollingen Series LXXXIX). Princeton, New Jersey: Princeton University Press.

Longino, Helen E. 1990. *Science as Social Knowledge: Values and Objectivity in Scientific Inquiry*. Princeton, New Jersey: Princeton University Press.

Lyotard, Jean-Francois. 1984. *The Post-Modern Condition: A Report on Knowledge*. Manchester: Manchester University Press.

Macleod, Roy. 1987. 'On Visiting the Moving Metropolis: Reflections on the Architecture of Imperial Science', in N. Reingold and M. Rothenburg (eds), *Scientific Colonialism: A Cross Cultural Comparison*. Washington: Smithsonian Institution Press, pp. 217–49.

Macleod, Roy and Dionne, Russel. 1979. 'Science and Policy in British India, 1858–1914: Perspectives on a Persisting Belief'. Proceedings of the Sixth European Conference of Modern South Asian Studies, Colloques Internationaux du CNRS, Asie du Sud: Traditions et Changements, CNRS, Paris.

Macleod, Roy and Kumar, Deepak (eds). 1995. *Technology and the Raj: Technical Transfers to India 1700–1947*. New Delhi: Sage Publications.

Mahadevan, T.M.P. 1969. 'Philosophical Trends vs History of Sciences in India: Orthodox Systems', *Indian Journal of History of Sciences* 1 and 2, pp. 27–41.

Marx, Karl. 1853. 'The British Rule in India', 25 June, and 'The Future Results of British Rule in India', 8 August, *New York Daily Tribune*.

—— 1977. *The Eighteenth Braumaire of Louis Bonaparte*. Moscow: Progress Publishers

Masini, Eleonara Barbieri. 1994. 'Science and Art in Perspective: Reflections in a Socio-historic Key', *World Futures* 40(1–3), pp. 45–8.

Mendelsohn, Everett. 1977. 'The Social Construction of Scientific Knowledge', in Everett Mendelsohn, Peter Weingart, and Richard Whitely (eds), *The Social Production of Scientific Knowledge*. Dordrecht: D. Reidel Publishing Company, pp. 3–26.

Mendelsohn, Everett. 1995. 'Science and the Construction of the Idea of Europe', *VEST* 4(8), pp. 59–64.

Merchant, Carolyn. 1980. *The Death of Nature: Women, Ecology and the Scientific Revolution*. San Francisco: Harper and Row.

Merton, Robert K. 1970. *Science, Technology and Society in Seventeenth Century England*. New York: Harper and Row.

Metcalf, Barbara. 1986. 'Hakim Ajmal Khan: Rais of Delhi and Muslim leader', in R. E. Frykenberg (ed.), *Delhi Through the Ages*. Oxford: Oxford University Press.

Mitter, Partha. 1994. *Art and Nationalism in Colonial India: 1850–1922*. Cambridge: Cambridge University Press.

Moore, Alvin Jr. and Coomaraswamy, Rama Poonambulam. 1988. *Selected Writings of Ananda Coomaraswamy*. New Delhi: Oxford University Press.

Murr, Sylvia. 1983. 'Les conditions d'emergence du discours sur l'Inde au siècle des lumières', *Collection Purusartha* 7, pp. 233–84.

—— 1986. 'Les Jésuites et l'Inde au XVIIIᵉ siècle: Praxis, utopie, préanthropologie', *Revue de l'Université d'Ottawa* 56(1), pp. 9–27.

Nader, Laura. 1996. 'Introduction: Anthropological Inquiry into Boundaries, Power and Knowledge', in Laura Nader (ed.), *Naked Science: An Anthropological Inquiry into Boundaries, Power and Knowledge*. London and New York: Routledge, pp. 1–25.

Nakayama, S. 1991. 'The Shifting Centres of Science', *Interdisciplinary Science Reviews* XVI(I), pp. 82–8.

Namboodripad, E.M.S. 1993. 'In Memory of Debiprasad: The Pioneer of Marxism in Indian Philosophy', *The Marxist*, April–June, pp. 1–8.

Nanda, M. 1991. 'Is Modern Science a Western Patriarchal Myth? A Critique of Populist Orthodoxy', *South Asia Bulletin* II(1/2), pp. 32–61.

Nandy, Ashis. 1980. *Alternative Sciences*. Delhi: Allied Publishers.

—— 1981. 'Counter-statement on Humanistic Temper', *Mainstream*, 10 October 1981, p. 16.

—— 1983. *The Intimate Enemy: Loss and Recovery of Self Under Colonialism*. New Delhi: Oxford University Press.

—— 1988. 'Introduction: Science as a Reason of State', in Ashis Nandy (ed.), *Science, Hegemony and Violence: A Requiem for Modernity*. New Delhi: Oxford University Press.

—— 1995. 'The Savage Freud: The First Non-Western Psychoanalyst and the Politics of Secret Selves in Colonial India', in *The Savage Freud and Other Essays on Possible and Retrievable Selves*. New Delhi: Oxford University Press.

Needham, Joseph. 1969. *The Grand Titration: Science and Society in East and West*. London: George Allen and Unwin.

—— 1973. 'The Historian of Science as Ecumenical Man', in Shigeru Nakayama and Nathan Sivin (eds), *Chinese Science: Explorations of an Ancient Tradition*. Cambridge, Massachusetts: The MIT Press, pp. 1–8.

—— 1977a. *Science and Civilization in China* (vol. 1). Cambridge: Cambridge University Press.

—— 1977b. *Science and Civilization in China* (assisted by Wang Ling) (vol. 2). Cambridge: Cambridge University Press (See Chapters 10 and 11 on the Tao Chai and Taoism, and the Mo Chia and the Ming Chia).

—— 1977c. *Science and Civilization in China* (assisted by Wang Ling and Kenneth Girdwood Robinson) (vol. 4). Cambridge: Cambridge University Press.

Nelson, Benjamin. 1981. *On the Roads to Modernity—Conscience, Science and Civilizations: Selected Papers by Benjamin Nelson* (edited by Toby E. Huff). Totowa, New Jersey: Rowman and Littlefield.

Pandey, G. 1992. *The Construction of Communalism in Colonial North India*. New Delhi: Oxford University Press.

Pannikar, K.N. 1980. 'Cultural Trends in Pre-colonial India: An Overview', *Studies in History* II(2), pp. 62–80.

Parusnikova, Zuzanna. (1992). 'Is a Post-modern Philosophy of Science Possible', *Studies in History and Philosophy of Science* 23(1), pp. 21–37.

Paul, Harry W. 1985. *From Knowledge to Power: The Rise of Science and Empire in France, 1860–1939*. Cambridge: Cambridge University Press.

Peiffer, Jeanne. (forthcoming). 'France', in J. Dauben and C.J. Scriba (eds), *Writing the History of Mathematics: Its Historical Development*. Birkhäuser–Basel: Science Network.

Pestre, Dominique. 1995. 'Pour Une Histoire Sociale et Culturelle des Sciences. Nouvelles Définitions, Nouveaux Objets, Nouvells Pratiques', *Annales* 50(3), pp. 487–522.

Petitjean, Patrick. 1992. 'Sciences et Empires: Un Thème Promèttéur Des Enjeux Cruciaux', in Patrick Petitjean, Catherine Jami, and Anne Marie Moulin, *Science and Empires*. Dordrecht: Kluwer, pp. 3–12.

Pickering, Andrew (ed.). 1992. *Science as Culture and Practice*. Chicago: Chicago University Press.

Polanco, X. 1985. 'Science in the Developing Countries: An Epistemological Approach on the Theory of Science in Context', *Quipu* 2(2), pp. 303–18.

Porter, Roy and Teich, Mikulas. 1992. 'Introduction' in Roy Porter and Mikulas Teich (eds), *The Scientific Revolution in National Context*. Cambridge: Cambridge University Press, pp. 1–10.

Potter, Karl H. 1977. *Encyclopedia of Indian Philosophies: Indian Metaphysics and Epistemology. The Tradition of the Nyaya-Vaisesika up to Gangesa*. New Delhi: Motilal Banarsidas.

Prakash, Gyan. 1990. 'Writing Post-Orientalist Histories of the Third World:

Perspectives from Indian Historiography', *Comparative Studies in Society and History* 32(2), pp. 383–408.

—— 1992. 'Science "Gone Native" in Colonial India', *Representations* 40, pp. 153–78.

Prathap, Gangan 1996. 'The Origins of Science; Part 1: Thales' Leap', *Resonance*, April, pp. 67-73.

Price, Derek J. de Solla. 1973. 'Joseph Needham and Science of China', in Shieguru Nakayama and Nathan Sivin (eds), *Chinese Science: Exploration of an Ancient Tradition.* Cambridge, Massachusetts: The MIT Press.

Pycior, Helena. 1983. 'Augustus De Morgan's Algebraic Work: The Three Stages', *Isis* LXXIV pp. 211–26.

Pyenson, Lewis. 1985. *Cultural Imperialism and Exact Sciences: German Expansion Overseas 1900-1930.* New York, Berne, Frankfurt am Main: Peter Lang.

—— 1989. 'What is the Good History of Science?', *History of Science* XXVII, pp. 353–89.

—— 1993a. 'Prerogatives of European Intellect: Historians of Science and the Promotion of Western Civilization', *History of Science* XXXI, pp. 289–15.

—— 1993b. *Civilizing Missions: Exact Sciences and French Overseas Expansion, 1830–1940.* Baltimore and London: Johns Hopkins University Press.

Qaiser, A.J. 1982. *The Indian Response to European Technology.* New Delhi: Oxford University Press.

Rahman, Abdur. 1982a. *Science and Technology in Medieval India: Bibliography of Source Materials in Sanskrit, Arabic and Persian.* Delhi: INSA.

—— 1982b. 'Science and Technology in Medieval India', in D.Chattopadhyaya (ed.), *Studies in the History of Science in India* (vol. 2), pp. 805–15.

—— 1984. 'A Conceptual Framework of History of Science in India', in A. Rahman (ed.), *Science and Technology in Indian Culture: A Historical Perspective.* NISTADS, pp. 21–43

—— 1987. *Maharaja Jai Singh II and Indian Renaissance.* New Delhi: Navrang.

—— 1996. 'Science and Social Movements: Bhakti and Sufi Movements, 10th–18th C'. Unpublished manuscript.

Raina, Dhruv. 1992. 'Mathematical Foundations of a Cultural Project or Ramchandra's Treatise through the Unsentimentalized Light of Mathematics', *Historia Mathematica* XIX, pp. 371–84.

—— 1998. 'Beyond the Diffusionist History of Colonial Science', *Social Epistemology* 12(2), pp. 203–13.

—— 1999. 'Nationalism, Institutional Science and the Politics of Knowledge: Ancient Indian Astronomy and Mathematics in the Landscape of French Enlightenment Historiography'. Institutionen för vetenskapsteori, Göteborgs Universitet, Rapport Nr. 201.

Raina, Dhruv and Habib, S. Irfan. 1990. 'Ramchandra's Treatise through the Haze of the Golden Sunset. An Aborted Pedagogy', *Social Studies of Science* 20, pp. 455–72.

Raina, Dhruv and Habib, S. Irfan. 1993. 'The Unfolding of an Engagement: The Dawn on Science, Technical Education and Industrialization', *Studies in History* 9(1), pp. 87–117.

—— 1995. 'Bhadralok Perceptions of Science, Technology and Cultural Nationalism', *Indian Economic and Social History Review* 32(1) , pp. 95–117.

—— 1996. 'The Moral Legitimation of Modern Science: Bhadralok Reflections on Theories of Evolution', *Social Studies of Science* 26(1), pp. 9–42.

—— 1999. 'The Missing Picture: The Non-emergence of a Needhamian History of Sciences of India', in S. Irfan Habib and Dhruv Raina (eds), *Situating the History of Sciences: Dialogues with Joseph Needham*. New Delhi: Oxford University Press, pp. 279–302.

Raina, Dhruv and Gupta, B.M. 1998. 'Four Aspects of the Institutionalization of Physics Research in India: Between Sociology and Bibliometrics', *Scientometrics* 42(1), pp. 17–40.

Raina, Dhruv and Jain, Ashok. 1997. 'From the Imperatives of Big Science to the Impoverishment of the University Research System in India', in John Krige and Dominique Pestre (eds), *Science in the Twentieth Century*. Amsterdam: Harwood Publishers.

Raina, Vinod, Chowdhury, Aditi, and Chowdhury, Sumit. 1997. *The Dispossessed: Victims of Development in Asia*. Hong Kong: Arena Press.

Raj, Kapil. 2000. '18th-century Pacific Voyages of Discovery and the Making of Europe', *History and Technology* 17(2), pp. 79–98.

Rashed, Roshdi. 1989. 'Problems of the Transmission of Greek Scientific Thought into Arabic: Examples from Mathematics and Optics', *History of Science* XXXVII, pp. 199–209.

—— 1994. *Development of Arabic Mathematics: Between Arithmetic and Algebra*. Dordrecht: Kluwer.

Ravetz, J.R. 1992. *The Merger of Knowledge with Power*. London/New York: Mansell Publishing.

Ramasubhhan, R. and Singh, B. 1987. 'The Orientation of the Public Sciences in a Post-colonial Society: The Experiences of India', in S. Blume, J. Blunders, L. Leydersdorff, and R. Whitley (eds), *The Social Direction of the Public Sciences: Sociology of Science Yearbook* (vol. XI). Dordrecht: D. Reidel, pp. 163–91.

Ray, J.N. 1961. 'Acharya Ray and Chemical Research in India', *Journal of the Indian Chemical Society* 38(8), pp. 423–31,

Ray, P. 1966. 'Prafulla Chandra Ray: 1861–1944', in *Biographical Memoirs of Fellows of National Institute of Science* (vol. I). New Delhi: Indian National Science Academy, pp. 58–76.

Ray, P.C. 1896a. 'On Mercurous Nitrate', *Journal of the Asiatic Society of Bengal* 65, pp. 1–9.

—— 1896b. 'Interaction of Mercurous Nitrite and Alkyl Iodides', *Proc. Chem. Soc. London* 12, p. 218.

—— 1896c, 'Mercurous Nitrite', *Zell. Anorg. Chem.* 12, p. 365.

—— 1902. *A History of Hindu Chemistry* (vol. I). Calcutta: Chuckervertty & Co. and Kegan Paul.

—— 1906. 'The Tantrists, the Rosicrucians and the Seekers After Truth', *The Modern Review* 16, pp. 237–39.

—— 1907. *A History of Hindu Chemistry* (vol. II). Calcutta: Chuckervertty & Co. and Kegan Paul

—— 1914. 'The Place of Mercury in the Periodic Table', *The Chemical News* CIX (2830), p. 85.

—— 1918. *Essays and Discourses*. Madras: G.A. Natesan & Co.

—— 1925. *Makers of Modern Chemistry*. Calcutta: Chuckervertty, Chatterjee & Co. and London: Probsthain & Co.

—— 1932. *Life and Experiences of a Bengali Chemist* Calcutta and London: Chuckerverity, Ghattesjee & Co. and Kegan Paul, Trubner.

Ray, P.C. and Dutta Bidhubhusan. 1911. 'Iwan Iwanoswitch Mendeleef', *The Modern Review* 21, pp. 460–62.

Reddy, A.K.N. 1978a. 'The Genetic Aspects of Western Technology', *Mazingira* 5.

—— 1978b. 'Technologies Appropriate to Rural Development'. INSA Seminar on Contribution of Science to Rural Development. New Delhi.

Restivo, Sal. 1983. *The Social Relations of Physics, Mysticism and Mathematics*. Dordrecht: D. Reidel.

—— 1994. 'The Theory Landscape in Science Studies: Sociological Traditions', in Sheila Jasanoff, Gerald E. Markle, James C. Petersen, and Trevor Pinch (eds), *Handbook of Science and Technology Studies*. London: Sage Publications, pp. 95–110.

Roşu, Arion. 1986. 'Marcelin Berthelot et L'Alchimie Indienne', *Bulletin de l'Ecole Franfaise d' Extreme-Orient* LXXV, pp. 67–78.

—— 1990, 'Marcelin Berthelot, Historien des Sciences', *Sudhoffs Archiv* 74(2), pp. 186–209.

Rouse, Joseph. (1987). *Knowledge and Power: Toward a Political Philosophy of Science*. Ithaca and London: Cornell University Press.

Roy, R.R. 1821. 'The Missionary and the Brahmin', *Brahmunical Magazine* 1.

Richards, Joan. 1995. 'The History of Mathematics and l'Espirit Humain: A Critical Appraisal', *Osiris*, 10, pp. 122–35.

Rushdie, S. 1991. *Imaginary Homelands*. London: Granta Books.

Russel, Colin. 1988. 'Rude and Disgraceful Beginnings: A View of History of Chemistry from the Nineteenth Century', *British Journal of History of Science* 21, pp. 272–94.

Sachau, Edward C. 1910. *Alberuni's India: An account of the religion, philosophy, literature, geography, chronology, customs, laws and astrology of India*. Delhi: Low Price Publications (reprinted in 1993).

Said, Edward W. 1978. *Orientalism*. London: Penguin.

—— 1990. 'Facts, Configurations, Transfigurations', *Race & Class* 32(1), pp. 1–16.

Said, Edward W. 1994. *Culture & Imperialism*. London: Vintage.

Salomon-Bayet, Claire (ed.). 1986. *Pasteur et la revolution pastorienne*. Paris: Payot.

Salomon, Jean-Jacques. 1992. *Le Destin Technologique*. Paris: Gallimard.

Salomon, Jean-Jacques and Lebeau, André. 1993. *Mirages of Development: Science and Technology for the Third World*. Boulder, Colorado: Lynne Reinner.

Sangwan, Satpal. 1985. 'European Impressions of Science and Technology in India', *Social Science Probings* 2(3), pp. 353–77.

Sardar, Ziauddin. 1993. 'Do Not Adjust Your Mind: Post-Modernism, Reality and the Other', *Futures*, October, pp. 877–93.

Sarkar, Benoy Kumar. 1946. *Education for Industrialization: An Analysis of Forty Years of Jadavpur College of Engineering and Technology (1905–45)*. Calcutta: National College of Education.

Sarkar, Sumit. 1975a. *The Swadeshi Movement in Bengal*. New Delhi: PPH.

—— 1975b. 'Rammohun and the Break with the Past', in V.C. Joshi (ed.), *Rammohun and the Process of Modernization in India*, New Delhi: Vikas, pp. 46–68.

Sarton, George. 1957. *The Study of the History of Science*. New York: Dover (first published by Harvard University Press in 1936).

Schott, Thomas. 1993a. 'The Movement of Science and of Scientific Knowledge: Joseph Ben-David's Contribution to its Understanding', *Minerva* XXXI (Winter), pp. 455–77.

—— 1993b. 'World Science: Globalization of Institutions and Participation', *Science, Technology and Human Values* XVIII (2), pp. 196–208.

Seal, B.N. 1915 (reprinted 1985). *The Positive Sciences of the Ancient Hindus*. Delhi: Motilal Banarsidas.

Sen, Amitabha. 1986. 'P.C. Ray's Contribution to Indian Science and Industry', in Santimay Chatterjee and Amitabha Sen (eds), *Acharya Prafulla Chandra Ray: Some Aspects of His Life and Work*. Calcutta: Indian Science News Association, pp. 31–77.

Sen, S.N. 1966. 'Changing Perspectives in the History of Sciences', *Science and Culture* 31(5), pp. 214–19.

—— 1988. 'The Character of the Introduction of Western Science in India during the Eighteenth and Nineteenth Centuries', *Indian Journal of History of Science* 1 and 2(23), pp. 112–22.

Seshadri, C.V. 1980. 'Energy in Indian Context', *Invited Lectures at Madurai Kamaraj University*.

—— 1982. 'Development and Thermodynamics: A Search for New Quality Markers'. *MSEPS*, II, MCRC, February.

Shapin, Steven. 1983. '"Nibbling at the Teats of Science": Edinburgh and the Diffusion of Science in the 1830s', in Ian Inkster and Jack Morrel (eds), *Metropolis and Province: Science in British Culture: 1780–1850*. Philadelphia: University of Pennsylvania Press, pp. 151–78.

—— 1989. 'The Invisible Technician', *American Scientist* 77, pp. 554–63.

—— 1992. 'Discipline and Bounding: The History and Sociology of Science as Seen through the Externalism-Internalism Debate', *History of Science* 30, pp. 334–69.

—— 1994. *A Social History of Truth: Civility and Science in Seventeenth Century England*. Chicago and London: University of Chicago Press.

—— 1996. *The Scientific Revolution*. Chicago: University of Chicago Press.

Shapin, Steven and Schaffer, Simon. 1985. *Leviathan and the Air-Pump: Hobbes, Boyle and the Experimental Life*. Princeton, New Jersey: Princeton University Press.

Shils, Edward. 1991. 'Reflections on Tradition, Centre and Periphery and the Universal Validity of Science: The Significance of the Life of S. Ramanujan', *Minerva* XXIX (Winter), pp. 391–419

Shiva, V. 1988. *Staying Alive: Women, Ecology And Survival In India*. New Delhi & London: Zed Books.

Shortland, Michael and Warwick, Andrew (eds). 1989. *Teaching the History of Science*. Oxford: Basil Blackwell.

Singh, A.N. 1936a. 'On the Use of Sines in Hindu Mathematics', *Osiris* 1, pp. 605–28.

—— 1936b. 'A Review of Hindu Mathematics upto the 12th Century', *Archeion* 18, pp. 43–62.

Singh, N. 1986. 'Linguistics and the Oral Tradition in the Period between the Decline of Harappan Culture and the Rise of Magadhan Culture', in Debiprasad Chattopadhyaya, *History of Science and Technology in Ancient India: The Beginnings*. Calcutta: Firma KLM, pp. 406–455.

Sismondo, Sergio. 1996. *Science without Myth: On Constructions, Reality and Social Knowledge*. Albany, New York: SUNY Press.

Snow, C.P. 1959. *The Two Cultures: And a Second Look*. Cambridge: Cambridge University Press.

Söderqvist, Thomas. 1997. *The Historiography of Contemporary Science and Technology*. Amsterdam: Harwood Academic Publishers.

Sörlin, Sverker. 1992. 'The International Contexts of Swedish Science: A Network Approach to the Internationalism of Science', *Science Studies* II, pp. 5–12.

Subbarayappa, B.V. 1967. 'An Estimate of the Vaisesika Sutra in the History of Science', *Indian Journal of History of Sciences* 2(1), pp. 22–24.

—— 1992. *In Pursuit of Excellence: A History of the Indian Institute of Science*. New Delhi: Tata Macgraw-Hill.

Storey, William K. 1996. 'Introduction', in William K. Storey (ed.), *Scientific Aspects of European Expansion—An Expanding World: The European Impact on World History 1450–1800* (Vol. 6). Adershot: Variorum, pp. xiii–xxi.

Thackray, Arnold. 1979. 'Natural Knowledge in Cultural Context: The Manchester Model', *American Historical Review* LXXIX, pp. 672–702.

Thackray, Arnold. 1980. 'History of Science', in P.T. Durbin (ed.), *A Guide to the Culture of Science, Technology and Medicine*. New York: The Free Press, pp. 3–69.

Thackray, Arnold and Merton, Robert K. 1972. 'On Discipline Building: The Paradoxes of George Sarton', *Isis* 63(219), pp. 473–95.

Thapar, Romila. 1993a. 'Durkheim and Weber on theories of societies and race relating to pre-colonial India', in Romila Thapar, *Interpreting Early India*. New Delhi: Oxford University Press, pp. 25–59.

―― 1993a 'The Contribution of D.D. Kosambi to Indology', in *Interpreting Early India*. New Delhi: Oxford University Press, pp. 89–113.

Thomas-Issac, T.M. and Ekbal, B. 1987. *Science for Social Revolution*. Thiruvananthapuram: Kerala Sastra Sahitya Parishad.

Toulmin, Stephen. 1990. *Cosmopolis: The Hidden Agenda of Modernity*. Chicago: University of Chicago Press.

Tucker, Aviezer. 1993. 'A Theory of Historiography as a Pre-science', *Studies in History and' Philosophy of Science* 24(4), pp. 633–67.

Turner, Stephen. 1990. 'Forms of Patronage', in Susan E. Cozzens and Thomas F. Gieryn (eds), *Theories of Science and Society*. Bloomington: Indiana University Press.

Uberoi, J.P.S. 1978. *Science and Culture*. New Delhi: Oxford University Press.

―― 1984. *The Other Mind of Europe: Goethe as a Scientist*. New Delhi: Oxford University Press.

Van den Daele, Wolfgang. 1977. 'The Social Construction of Science: Institutionalisation and Definition of Positive Science in the Latter Half of the Seventeenth Century', in Everett Mendelsohn, Peter Weingart, and Richard Whitely (eds), *The Social Production of Scientific Knowledge*. Dordrecht: D. Reidel, pp. 3–26.

Vincenti, Walter G. 1990. *What Engineers Know and How they Know it*. Baltimore: Johns Hopkins University Press.

Visvanathan, Shiv. 1985. *Organizing for Science*. Oxford: Oxford University Press.

―― 1992. 'Reinventing Gandhi', *IUMDA Newsletter* 5, pp. 34–64.

―― 1992. 'Footnotes to Vavilov: An Essay on Gene Diversity'. New Delhi: Centre for Studies of Developing Societies (mimeograph).

―― 1997. *A Carnival for Science: Essays on Science, Technology and Development*. New Delhi: Oxford University Press.

Weingart, Peter. 1993. 'Science Abused? Challenging a Legend', *Science in Context* VI(2), pp. 555–67.

Will, Pierre-Étienne. 1995. 'Modernisation Less Science? Some Reflections on China and Japan before Westernisation', in Hashimoto Keizô, Catherine Jami, and Lowell Skar (eds), *East Asian Science: Tradition and Beyond*. Osaka: Kansai University Press, pp. 33–48.

Whewell, William. 1857. *History of the Inductive Sciences, from the Earliest to the Present Time* (3 vols). London: Parker.

Wittrock, Björn. 'Early Modernities: Varieties and Transitions', *Daedalus*, 127(3), pp. 190-40.

Windelband, Wilhelm. 1958. *A History of Philosophy* (2 vols). New York: Harper and Brothers (first published in 1893).

Werskey, Gary. 1978. *The Visible College*. New York: Holt, Reinhart and Winston.

Wertheim, Margaret. 1997. *Pythagoras' Trousers: God Physics and the Gender Wars*. London: Fourth Estate.

Young, Robert. 1990. *White Mythologies: Writing History and the West*. London and New York: Routledge.

Zachariah, Mathew and Sooryamoorthy, R. 1994. *Science for Social Revolution? Achievement and Dilemmas of a Development Movement: The Kerala Sastra Sahitya Parishad*. New Delhi: Vistaar.

Zilsel, Edgar. 1942. 'The Sociological Roots of Science', *American Journal of Sociology* 47, pp. 544–62.

Index

Gould, Stephen Jay, 180
Graham, Loren, 3, 43
Greek Science, 91
growth, Rostowian model of, 16
Gruber, Howard, 73
Guay, Y., 77
Guenon, René, 95–6
Guillemain, Bernard, 64, 140
Gupta, B.M., 188

Habib, S. Irfan, 4, 24, 29–30, 34,
    41–3, 51, 53–4, 97, 123–5, 141,
    152–3, 163–5, 183
Hacking, Ian, 190
Hagendijk, Rob, 174
Haldane, J.B.S., 29, 151
Haraway, Donna, 200
Harding, Sandra, 4, 10–12, 154,
    185, 197, 200, 202
Hardy, G.H., 160, 163, 165–8, 173
Hartog, 65
Harun, Khalif, 69
Hasse, Raimund, 173
Hayles, Katherine, 88
Headrick, D.R., 176
Hegel, G.W.F., 7
Hess, David J., 3
Hessen, Boris, 3
Hiebert, Edwin N., 71
Hindu sciences, 39
Hinduism, 88–9, 93
Histoire de la Chimie, 66
historiography,
    Big Picture, 15
    colonial, 23
    contextualization, 8
    definition, 9
    evolution, 19
    and historians of science, 192,
        196
    internal and external, 13–14
    liberal, 30
    Marxist, 152
    of modernity, 21
    modernization, 17

nationalist, 120
and Needham, 143–6
orientalist, 14
of science, 6–9, 26, 59–70, 85, 91,
    117–18, 131, 148, 150, 153,
    192, 194, 196
of scientific revolution, 15, 17,
    147, 150, 194, 200–01
within social theory of science,
    6–9
history,
    art of, 88–92
    notion of, 24
    politics and object of, 98–101
History of Chemistry, 66
History of Greek Alchemy, 169
History of Hindu Chemistry, 56, 60,
    65–6, 71, 78, 107, 170
History of Indian and Indonesian Art,
    86
History of Indian Philosophy, 126
history of science,
    Basalla model, 180–3
    demands, 133–6
    as a discipline, 83, 192
    elements of diffusionism, 183–6,
        188–90
    Eurocentrism, 124, 197, 199
    future trajectories, 192–203
    hegemonic conception, 2
    historiography of internal and
        external, 85
    history, 146
    ideological prejudice, 153
    in India (1966–94), 105–36
    institutional barriers, 133–6
    lost opportunity, 127–31
    Marxist externalist, 5
    Needhamian, 139–56, 178–9
    and political colonization,
        186–8
    positivist and triumphalist, 1, 4
    professional identity, 83, 85, 108
    research, 147
    science, scientists and, 105–36